AF843015

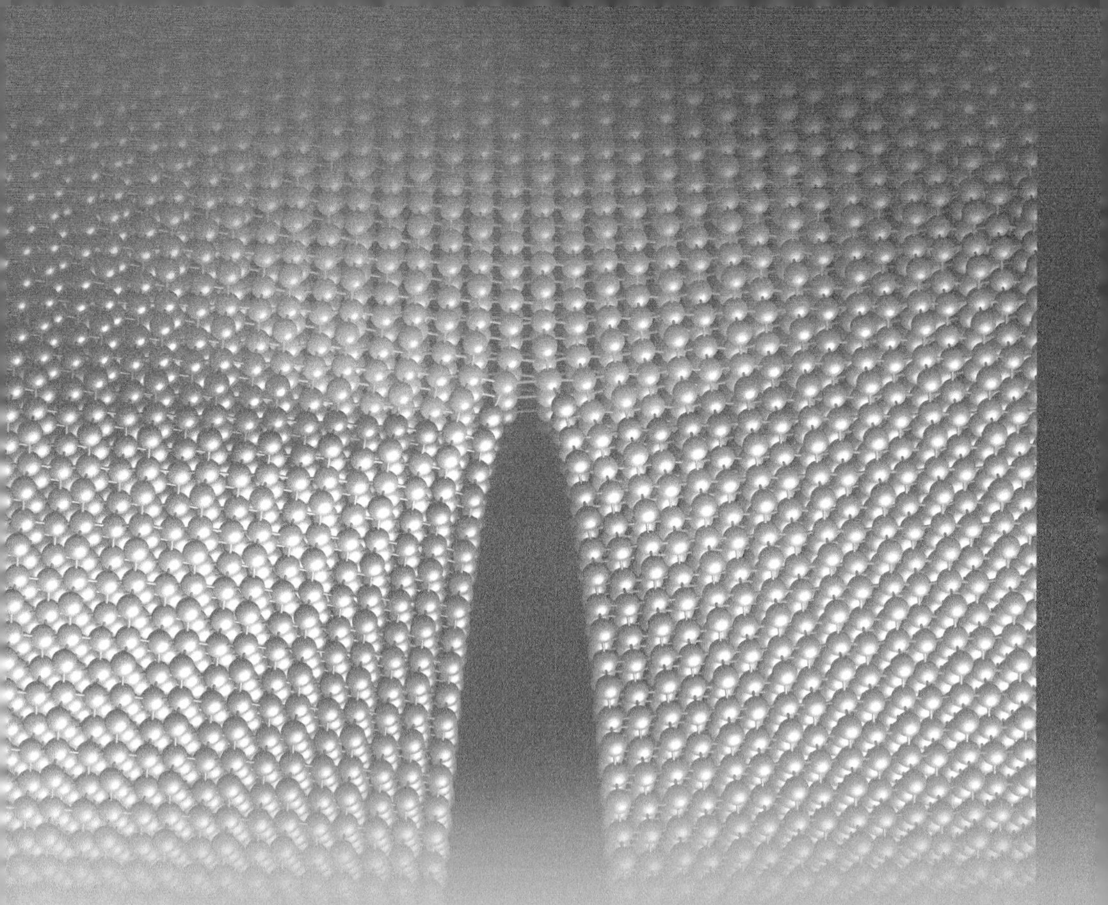

High Temperature Fracture Mechanics

High Temperature Fracture Mechanics

Takayuki Kitamura

JENNY STANFORD
PUBLISHING

Published by

Jenny Stanford Publishing Pte. Ltd.
101 Thomson Road
#06-01, United Square
Singapore 307591

Email: editorial@jennystanford.com
Web: www.jennystanford.com

British Library Cataloguing-in-Publication Data
A catalogue record for this book is available from the British Library.

High Temperature Fracture Mechanics

Copyright © 2025 Jenny Stanford Publishing Pte. Ltd.
All rights reserved. This book, or parts thereof, may not be reproduced in any form or by any means, electronic or mechanical, including photocopying, recording or any information storage and retrieval system now known or to be invented, without written permission from the publisher.

For photocopying of material in this volume, please pay a copying fee through the Copyright Clearance Center, Inc., 222 Rosewood Drive, Danvers, MA 01923, USA. In this case permission to photocopy is not required from the publisher.

ISBN 978-981-5129-55-7 (Hardcover)
ISBN 978-1-003-65328-8 (eBook)

Contents

Preface

High temperature reliability is a critical issue for propulsion equipment, such as jet and rocket engines, and power-generation equipment, such as gas turbines, as well as electronic devices and sensors as their materials are exposed to severe conditions during service. This means that fracture prevention is an urgent challenge and a key factor in overcoming weaknesses in future equipment design.

It is difficult to elucidate the high temperature strength of materials due to the synergistic effects of creep damage progression with time caused by thermal activation and fatigue damage caused by cyclic loading. Thus, a systematic understanding of these phenomena based on solid mechanics remains underdeveloped in the engineering community. In particular, since the propagation of small cracks, which are initiated in the early stage of life, dominates the fracture of a material, elucidating the mechanics of crack propagation leads to an understanding of the ultimate failure of structural components. This book systematically explains the mechanics governing crack propagation.

Since there are relatively well-established laws for the "deformation" of a material, the rapid development of computers has enabled highly accurate predictions. However, in the simulation of "fracture," it is necessary to introduce empirical formulas tailored for specific phenomena. For failure at room temperature such as fracture toughness and fatigue, the mechanics of crack propagation, known as fracture mechanics, has been well developed, providing a robust framework for accurately and efficiently predicting fracture behavior, and numerous comprehensive textbooks have been published on such general fracture mechanics.

In contrast, fracture mechanics related to high temperature fatigue is characterized by strong nonlinearity and time-dependent effects, making it inherently distinct. Consequently, there are no systematic and accessible textbooks available on this subject at present. This book meticulously explains these difficulties step by step from the basics and presents the fracture mechanics under complex creep–

fatigue interaction. It also addresses the mechanics of small crack propagation, which is strongly affected by the microstructure of the material. By leveraging this knowledge, computational simulations can be effectively utilized to predict "high temperature fracture" in a rational and reliable manner.

I hope this book will be helpful for researchers and engineers to understand high temperature fracture mechanics.

Takayuki Kitamura

March 2025

Chapter 1

Introduction

1.1 Diversity and Complexity of Material Strength at High Temperatures

Researchers and engineers often express frustration over the difficulty of understanding high temperature strength. The complexity of material strength at high temperatures arises from diverse phenomena, which are exemplified by numerous terms assigned based on stress, strain, and temperature. Beyond the phenomena pertinent to room temperature material strength such as tensile strength and fatigue, there are various terms specific to high temperature conditions. For instance, creep-related terms include static creep, dynamic creep, cyclic creep, cyclic-temperature creep, and so on. Fatigue-related terms encompass creep fatigue, thermal fatigue, thermo-mechanical fatigue, and so on. Additionally, phenomena like relaxation and elastic follow-up further illustrate the complexity. This complexity is a significant reason why comprehensive textbooks systematically explaining fracture mechanics at high temperatures are scarce. One of the objectives of this book is to provide a comprehensive understanding of these various phenomena through the lens of fracture mechanics, specifically focusing on crack tip failure mechanics.

High Temperature Fracture Mechanics
Takayuki Kitamura
Copyright © 2025 Jenny Stanford Publishing Pte. Ltd.
ISBN 978-981-5129-55-7 (Hardcover), 978-1-003-65328-8 (eBook)
www.jennystanford.com

The investigation of high temperature strength has a history exceeding a century, and that of high temperature fracture mechanics alone spans well over fifty years. While the contents of this book are grounded in the pioneering work of previous researchers, its primary aim is not to review this extensive history but to provide the essence of concept and mechanics.

1.2 Time-Dependent Deformation and Fracture

The first key factor of complexity stems from creep, which brings about "time dependence" on deformation and fracture, and markedly differentiates high temperature strength from room temperature one. For instance, in addressing deformation, activation of crystal defects (such as vacancies and dislocations) contributes to a variety of creep mechanisms through phenomena including atomic diffusion, climbing of dislocation, grain boundary sliding, and so on that progress over time. Furthermore, these mechanisms vary significantly with temperature/load conditions and are greatly influenced by cyclic loading. As a result, the time-dependent change in strain emerges in a complex manner. Additionally, even in the simplest form of creep, the deformation and fracture exhibit strong non-linearity with respect to the relationship between stress and strain (strain rate). Of course, considering crack propagation, the focus is on the singular fields near the crack tip. Thus, in high temperature fracture mechanics, the exploration of "time dependent" and "non-linear" singular fields forms a fundamental (Chapters 3 and 4).

At high temperatures, environmental influences such as oxidation can also impact material strength. Since the influence also exhibits time dependence, it is sometime challenging to distinguish it from the creep. The primary difference is that the environmental influence shows the time dependence in fracture at the crack tip, while creep has the time dependence in not only fracture but also deformation characteristics. This book is dedicated to meticulously pursuing fracture mechanics that address the time dependence due to creep. Conversely, when time dependence that cannot be addressed by the fracture mechanics explained in this book becomes prominent, it is necessary to consider environmental factors.

1.3 Interaction between Creep and Fatigue

The second key factor of complexity in high temperature strength is caused by interaction between creep and fatigue, a phenomenon where creep and fatigue intricately influence each other. The former is characterized by the time dependence, while the latter is featured by the stress-cycle dependence. One of the highlights of this book is detailed explanation of mechanics and mechanism of crack propagation under such creep–fatigue interaction (Chapters 5 to 8). It is well-known that the failure process is markedly accelerated by the interaction due to mixture of cyclic damage and creep damage, rather than being a simple sum of each damage.

In short, high temperature fracture mechanics is characterized by the following factors:

(a) Time-dependent effect
(b) Non-linear effect
(c) Cyclic loading effect

1.4 Mechanics of Fracture Driving Force and Fracture Resistance

Fracture mechanics consists of two stages. The first stage involves identifying mechanical quantity that captures the stress singular field near the crack tip. This means an issue of "deformation" analysis concerning the singular fields of above (a), (b), and (c). Considering that stable crack growth arises from local fracture at the crack tip, the governing quantity of singular field represents "fracture driving force."

Merely determining stress and strain (strain rate) state fields near the crack tip does not suffice to understand crack propagation phenomena. The second stage involves mechanics of the "fracture" amount at the crack tip under given fracture driving force. Materials have their own peculiar resistance for the fracture, depending on the temperature/load conditions, namely "fracture resistance." Difficulty to explore fracture resistance arises from the difference of fracture mechanisms by (a), (b), and (c).

The scarcity of comprehensive literature on high temperature fracture mechanics is due to the intricate influence of creep–fatigue interactions on both across (a), (b), and (c). By identifying "fracture driving force" and "fracture resistance," a universal fracture mechanics law can be derived. The complexity from the viewpoint of these has been explained in detail in this book.

1.5 Fracture Mechanics of Microstructurally Small Cracks

One major unresolved challenge in the field of high temperature fracture mechanics lies in the influence of microstructures. Traditional fracture mechanics mainly deals with homogeneous materials. Many metallic/ceramic materials consist of polycrystalline structures composed of countless tiny grains. When a crack is significantly larger than the grain size, the homogenized approximation holds, making fracture mechanics effective. However, in high temperature fatigue of materials (*e.g.*, heat-resisting alloys) without pre-existing cracks, tiny cracks on the scale of microstructures form and propagate, leading to failure of material. The fracture mechanics governing such propagation remains to be established. Additionally, some alloys possess diverse microstructural features, including those composed of large grains, ones with directional grains such as unidirectionally solidified alloys, and even single-crystal alloys with finer structures like dendrites. In all these cases, the influence of microstructures poses a significant challenge.

One distinctive feature of this work is its description of the propagation of microstructurally small cracks based on the knowledge of high temperature fracture mechanics, serving as one of the future developmental challenges (Chapter 9).

1.6 Papers Published in Non-English

In Japan, since the 1970s, several prominent research groups have actively advanced studies on creep and creep–fatigue crack propagation based on non-linear fracture mechanics. Due to the circumstances of the time, many of investigation reports were written in Japanese and have remained unknown within the global framework of high temperature fracture mechanics until now. However, these works can be freely accessed online, it is now possible to read these studies in English (and other languages) with the recent advancements in translation software. This book features a proactive introduction to fracture mechanics based on these systematic Japanese studies, including URLs for reference. Although I have not examined the original texts in multiple languages and may have overlooked significant results, I believe this approach is crucial for the future development of comprehensive specialized books.

1.7 Future Directions

This book aims to provide the foundational principles of high temperature fracture mechanics, contributing a crucial part of the academic groundwork for emerging challenges. It is evident to possess the relevance to the strength and reliability of heat-resisting materials in new energy equipment utilizing environmentally friendly technologies. Furthermore, its importance is increasingly recognized in the context of nano/micro joints (low melting temperature metals) in extreme environments.

However, it is important to emphasize that with the advancement of science, research into new material processing and usage is progressing. Incorporating high temperature strength studies from this book into these areas will become necessary in the near future. Recent trends simply reveal potential developments such as the following:

(a) Multiphysics

While diffusion induced by stress is one of the major factors in creep, electromigration is another diffusion phenomenon driven by electric current. This is a typical example of fracture stemming from multiphysics, resulting in failure of electronic devices.

Recent investigation reveals that mechanical loading combined with other physical stimuli can induce characteristic time-dependent deformation. For example, it has been reported that applying high voltage can make bulk ceramics exhibit ductile creep behavior under a load at relatively low temperatures. Furthermore, it is reported that blocking visible light can induce significant plastic deformation in semiconductors, which is another example. Thus, multiphysics, where deformation occurs due to the combination of load and other physical stimuli, is gaining attention.

(b) Nanomaterials

In Coble creep, which involves grain boundary diffusion in polycrystals, the deformation rate is inversely proportional to the cube of the grain size. Hence, creep deformation becomes prominent in small components or microstructures. Simultaneously, nanomaterials exhibit unique properties, leading to active research and development as advanced materials. In the future, reliability under creep–fatigue conditions is expected to become a significant challenge.

(c) Complex Structured Materials

The ability to fabricate structures using 3D printing has dramatically increased the freedom of microstructural shape-design. It is now possible to incorporate crystallographic design within these structural materials, creating nested muti-scale structures. Additionally, multimaterial structures, combining different materials, have become more feasible. Addressing high temperature fracture mechanics for these advanced structures is expected to be a future challenge.

1.8 Structure of This Book

Chapter 2 explains the fundamentals of fracture mechanics for non-linear elastic bodies focusing on the concept of J-integral, which provides the foundation for understanding fracture mechanics at high temperatures.

The mechanical behavior at high temperatures is characterized by creep, which brings about the time dependence on deformation/fracture along with the non-linearity in the stress–strain rate relationship. Chapter 3 explains the fundamentals of creep mechanics and demonstrates the characteristics of crack propagation in static creep.

To advance the understanding of the mechanics of creep–fatigue interaction, it is necessary to investigate the influence of stress change on creep behavior. Chapter 4 presents the experimental results in creep with load change and explains the transient behavior of fracture driving force after load change based on analytical research in terms of material factor and mechanical one.

The coexistence of two different characteristics such as fatigue and creep, which respectively bring about cycle dependence and time dependence on crack propagation, is the root cause of the difficulty in understanding crack propagation behavior in high temperature fatigue. From Chapters 5 to 7, this complexity is explained step by step. Chapter 5 is dedicated to explaining the distinct features of these two fundamental mechanisms/mechanics.

The purpose of Chapter 6 is to extend the concept of transient behavior of creep crack propagation as shown in Chapter 4 to cracking under cyclic load conditions, such as time-dependent fatigue (creep-dominated fatigue), and examine the ability of the creep fracture mechanics through experiments.

Chapter 7 analyzes mechanics of crack propagation in creep–fatigue interaction. Unified fracture mechanics laws in high temperature fatigue are derived based on series of experimental results, and this can provide a comprehensive understanding of phenomena that are referred to by various names under different load/temperature conditions. The applicability for thermal fatigue, which is one of the most complex phenomena of high temperature fatigue, is presented.

There remain fundamental questions regarding crack propagation under creep–fatigue interaction unanswered. Chapter 8 explains the complexity and difficulty of these, which are not fully addressed in Chapter 7, and outlines the main points expected for future research.

The fracture mechanics concept described up to Chapter 8 deals with propagation of large cracks, where the crack length is significantly larger than the size of the microstructure. Cracks initiated on the smooth surface of components are comparable in size to the microstructure, and their propagation leads to fracture in creep and fatigue at high temperatures. Chapter 9 explains the fundamental behavior of such microstructurally small cracks in terms of the fracture mechanics.

Chapter 2

Fundamentals of Non-linear Fracture Mechanics

2.1 Linear Fracture Mechanics

2.1.1 Stress and Strain Field Near Crack Tip

Consider an area near a crack in an isotropic linear elastic plate as schematically depicted in Fig. 2.1. The asymptotic stress field near the crack tip takes the following form using the polar coordinates (r, θ) with the crack tip as the origin [Williams, 1957]:

$$\sigma_{ij} = \frac{K_I}{\sqrt{2\pi r}} F_I(\theta) + \frac{K_{II}}{\sqrt{2\pi r}} F_{II}(\theta) + \frac{K_{III}}{\sqrt{2\pi r}} F_{III}(\theta) \tag{2.1}$$

where $F_I(\theta)$, $F_{II}(\theta)$ and $F_{III}(\theta)$ are the known functions of θ, and K_I, K_{II}, and K_{III} are the linear elastic stress intensity factors, respectively.

This equation presents following important features:

(1) It holds regardless of boundary conditions such as shape and size of body and external loads.

(2) The stress field can be separated into three modes. As illustrated in Fig. 2.2, each term on the right side represents (1st term: Fig. 2.2(a)) the stress field of crack tip opening,

High Temperature Fracture Mechanics
Takayuki Kitamura
Copyright © 2025 Jenny Stanford Publishing Pte. Ltd.
ISBN 978-981-5129-55-7 (Hardcover), 978-1-003-65328-8 (eBook)
www.jennystanford.com

(2nd term: Fig. 2.2(b)) the one of in-plane crack tip shearing, and (3rd term: Fig. 2.2(c)) the one of out-of-plane crack tip shearing, namely Mode I, Mode II, and Mode III, respectively.

(3) At the crack tip, $r = 0$, stress and strain are infinite, indicating singularity with the exponent of $r^{-1/2}$ for all modes. Each mode has its own independent singularity.

(4) The intensity of the singular field is uniquely expressed by the linear elastic stress intensity factor in each mode.

The strain field, ε_{ij}, possesses a similar form with the singularity of $r^{-1/2}$.

Unless otherwise specified, this book will focus on Mode I, which has the greatest impact on the fracture of components. Readers can easily envision concrete examples of pure Mode I through the illustration in Fig. 2.3, cracking in tension (left) and bending (right).

The asymptotic stress near the crack tip in Mode I, the first term of Eq. (2.1), is formulated in detail as follows:

$$\begin{Bmatrix} \sigma_{11} \\ \sigma_{22} \\ \sigma_{12} \end{Bmatrix} = \frac{K_I}{\sqrt{2\pi r}} \cos\frac{\theta}{2} \begin{Bmatrix} 1 - \sin\frac{\theta}{2}\sin\frac{3\theta}{2} \\ 1 + \sin\frac{\theta}{2}\sin\frac{3\theta}{2} \\ \sin\frac{\theta}{2}\cos\frac{3\theta}{2} \end{Bmatrix} \tag{2.2}$$

Under Mode I conditions, the influence of external load and shape/size of components is all encompassed in the value of K_I. This suggests that K_I is a candidate governing parameter for a local event at the crack tip, such as local fracture meaning crack propagation. This also implies that the characteristics of K_I evaluated in laboratory crack propagation experiments using a standard test piece under standard loading can be directly applied to crack propagation in actual components.

When applied to actual materials, non-linear deformation such as plasticity may occur in high-stress regions near the crack tip. Even in such cases, if the plastic deformation is confined in a small region near the crack tip (small-scale yielding), it is possible to use the linear elastic intensity factor as the governing mechanical parameter in crack propagation.

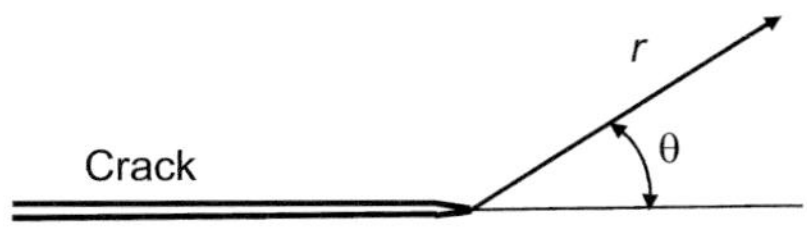

Figure 2.1 A crack in a linear elastic body.

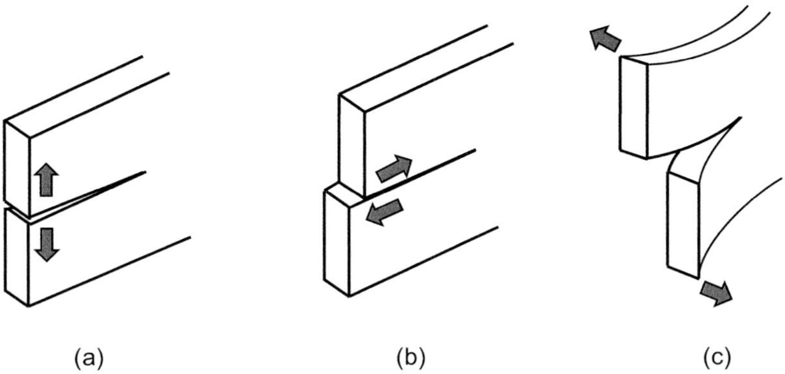

Figure 2.2 Cracking modes. (a) Mode I, opening mode, (b) Mode II, in-plane shear mode, and (c) Mode III, out-of-plane shear mode.

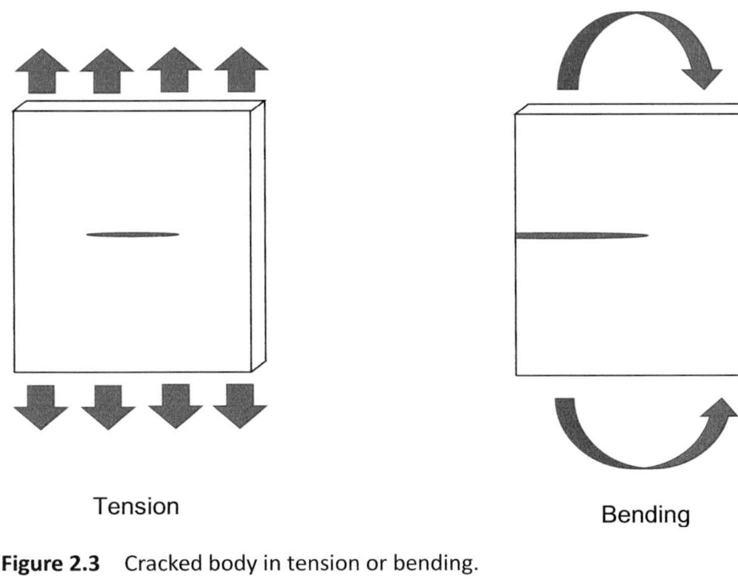

Figure 2.3 Cracked body in tension or bending.

2.1.2 Energy Release Rate

Consider a cracked body loaded by a device. The potential energy of the entire system, Π, is the summation of the strain energy of the cracked body and the energy stored in the device. If the crack does not propagate, the change in potential energy ΔΠ is zero because even if the load is changed, energy simply moves back and forth between the cracked body and the loading device. On the other hand, when the crack propagates by a length of Δa, the potential energy of the system decreases. The change ΔΠ with respect to the infinitesimal crack growth $\Delta a \to 0$ is defined as the energy release rate [Irwin, 1956].

$$G = -\frac{\partial \Pi}{\partial a} \tag{2.3}$$

When a crack in a body with a length of a subjected to a load, P, extends to $a + \Delta a$, the change in strain energy stored in the body, ΔW, is given by

$$\Delta W = W(a+\Delta a) - W(a) = \frac{P\Delta\delta}{2} = S_a \tag{2.4}$$

using their loading curves (δ is displacement) as depicted in Fig. 2.4(a). On the other hand, since the load is constant, the energy released from the loading device is $P\Delta\delta$ (It should be noted that this is negative as stored energy in the loading device). Therefore, the change, ΔΠ, in the potential energy of the entire system, Π, is given by

$$\Delta\Pi = \frac{P\delta\Delta}{2} - P\Delta\delta = -\frac{P\Delta\delta}{2} = -S_a \tag{2.5}$$

$$G = \lim_{\Delta a \to 0} S_a \tag{2.6}$$

When a cracked body is loaded by displacement δ, the change in strain energy, ΔW, due to crack extension is the area (negative) of S_b in Fig. 2.4(b). Since no work is done from the device during the crack extension, the change, ΔΠ, in the potential energy of the entire system, Π, is given by

$$\Delta\Pi = -\frac{\delta\Delta P}{2} - 0 = -\frac{\delta\Delta P}{2} = -S_b \tag{2.7}$$

$$G = \lim_{\Delta a \to 0} S_b \tag{2.8}$$

In general loading case, the potential energy change lies somewhere in between Figs. 2.4(a) and (b). Here, considering the crack extension of $\Delta a \to 0$, the difference between S_a and S_b is a second-order infinitesimal quantity, so it can be ignored compared to S_a and S_b ($S_a \cong S_b$). Thus, the energy released due to infinitesimal crack extension is uniquely determined.

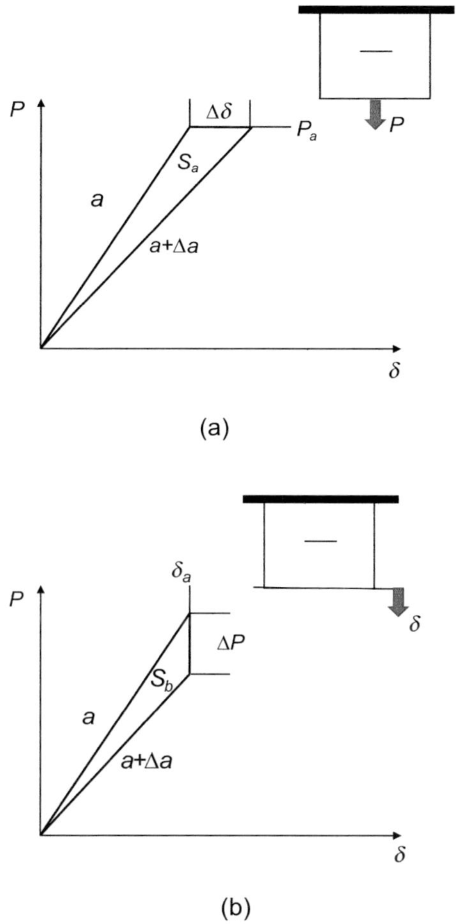

Figure 2.4 Load-displacement curves in linear elastic body with a crack of *a* or *a* + Δ*a* (a) under load-control and (b) under displacement-control.

It is proved that G has the following relation with K_I [Irwin, 1957],

$$G = \frac{K_I^{\,2}}{E'} \tag{2.9}$$

$$E' = E \qquad \text{in plane stress}$$

$$E' = \frac{E}{1-\nu^2} \qquad \text{in plane strain}$$

where E is Young's modulus and ν is Poisson's ratio.

2.2 J-integral in Non-linear Elastic Body

2.2.1 Definition and Features of J-Integral

For elastic bodies including non-linear elastic ones, the J-integral is defined as the following line integral for a path Γ surrounding the crack tip from the lower crack surface to the upper one as illustrated in Fig. 2.5 [Rice, 1968a].

$$J = \int_{\Gamma} (Wn_1 - T_i \frac{\partial u_i}{\partial x_1}) ds \tag{2.10}$$

where n_1 is x_1-component of outward normal vector on the integration path Γ, T_i is traction, and u_i, is the component of displacement vector. W is the strain energy density given by

$$W = \int_0^{\varepsilon_{ij}} \sigma_{ij} d\epsilon_{ij} \tag{2.11}$$

The J-integral has the following important features:

(1) The value does not depend on the integral path (path independence) [Rice, 1968a] (see Appendix 2-A).
(2) It has physical meaning of energy release rate [Rice, 1968b] (see Appendix 2-B)

$$J = G \tag{2.12}$$

These features hold true regardless of the stress–strain relation (constitutive equation). Of course, the J-integral can be defined as the energy release rate at first and can derive the line-integral, inversely.

Because of the feature (2), J in a linear elastic body possesses the following relation with K_I

$$J = \frac{K_I^{\ 2}}{E'} \tag{2.13}$$

In the case of unreduced stress, a plastic body in total strain theory superficially behaves the same as a non-linear elastic body. Therefore, the properties of the J-integral described above can be formally applied. Conversely, it should be noted that the J-integral cannot be applied when stress reduces in a plastic body.

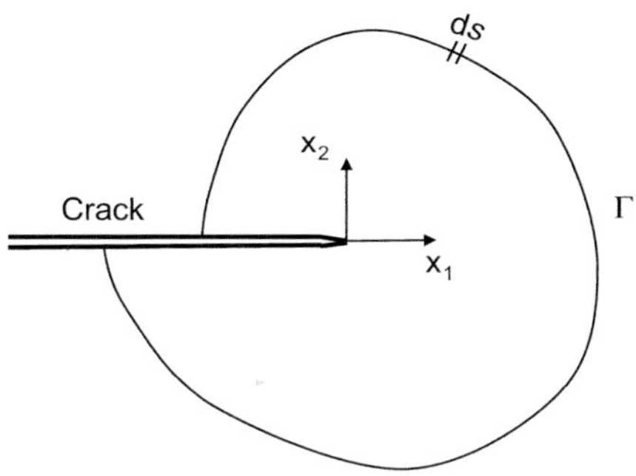

Figure 2.5 Integral path around a crack tip for the J-integral.

2.2.2 Singular Stress Field Near the Crack Tip

When the stress–strain relation follows a power law,

$$\epsilon = B_P \sigma^{n_P} \tag{2.14}$$

where B_P and n_P are material constants, the J-integral represents the intensity of the stress and strain singular fields near the crack tip. These are called HRR singularity and are expressed by the following equations (Appendix 2-C) [Hutchinson, 1968; Rice and Rosengren, 1968]:

$$\sigma_{ij} = \left(\frac{J}{I_{n_p} B_p r} \right)^{\frac{1}{n_p+1}} \tilde{\sigma}_{ij}(\theta) \tag{2.15}$$

$$\varepsilon_{ij} = \left(\frac{J}{I_{n_p} B_p r} \right)^{\frac{n_p}{n_p+1}} \tilde{\varepsilon}_{ij}(\theta) \tag{2.16}$$

where I_n is a function of n_p, $\tilde{\sigma}_{ij}(\theta)$, and $\tilde{\varepsilon}_{ij}(\theta)$ are functions of θ.

This presents the following features:

(1) These hold near the crack tip regardless of boundary conditions, *e.g.*, shape and size of body and external load.
(2) Stress and strain are infinite at the crack tip, indicating the singularities with the exponent of $r^{-1/(n_p+1)}$ and $r^{-n_p/(n_p+1)}$, respectively.
(3) The intensity of the singular fields is uniquely represented by the J-integral, J.

These suggest that the stable fracture near the crack tip, namely crack propagation, is governed by the J-integral. Although the derivation of these relationships requires some bold assumptions unlike the derivation of Eq. (2.2) for a linear elastic body, these have a similar meaning in crack propagation phenomena. Thus, this is understood as an extension of linear elastic fracture mechanics toward non-linear one. Of course, the features hold true for a plastic body without reduced stress condition, so the J-integral is applicable to a crack in a plastic body with limitation of loading.

2.2.3 Experimental Evaluation Method of J-Integral

The magnitude of J-integral in a plastic body can be readily evaluated by computational methods, such as finite element analysis. However, there are some troublesome points, such as a precise stress–strain relation (constitutive equation) needed. Then, there are simple methods developed for evaluation at the laboratory level, as explained below.

For a plate specimen with a short ligament (non-cracked part ahead of the crack tip—the length between the crack tip and side surface), commonly referred to as a deep crack, an experimental evaluation method of the J-integral magnitude is proposed using the loading curve [Rice *et al.*, 1973]. To prevent confusion in the subsequent explanation, the equations in this section include the plate thickness, h_t, even though the expressions in other parts of this book are based on a two-dimensional form, where $h_t = 1$ (see Appendix 2-D).

(1) Plate specimen with a single-edge crack under bending (right

in Fig. 2.3)

The linear and non-linear parts are usually evaluated separately, and the J-integral is often given as the sum of both.

$$\mathrm{J} = \frac{K^2}{E} + \frac{2S_p}{h_t b} \tag{2.17}$$

When the non-linearity dominates, it becomes

$$\mathrm{J} = \frac{2S_p}{h_t b} \tag{2.18}$$

where S_p is the area in the P–δ curve shown in Fig. 2.6(a).

(2) Plate specimen with a center crack under tension (left in Fig. 2.3)

The J-integral is evaluated by

$$\mathrm{J} = \frac{K^2}{E} + \frac{S_p}{h_t b} \tag{2.19}$$

where S_p is the area in the P–δ curve shown in Fig. 2.6(b).

A method without the loading history (P–δ curve) is proposed by applying the power law, Eq. (2.14) [Ohji *et al.*, 1978]. For a plate specimen with a center crack under tension, it is given by

$$\mathrm{J} = \frac{n_p - 1}{n_p + 1}\frac{P}{h_t b}\delta \tag{2.20}$$

Furthermore, replacing the load point displacement, δ, with the crack center opening displacement, V, it is reasonably applicable to a wide range of crack depth (ligament). Since n_p is usually large in comparison with unity, the evaluation equation becomes [Ohji *et al.*, 1978]

$$\mathrm{J} = \frac{n_p - 1}{n_p + 1}\frac{P}{h_t b}V \cong \frac{P}{h_t b}V = \sigma_{net}V \tag{2.21}$$

Based on the finite element analysis, the J-integral for a minute crack (or a crack in an infinite plate) under remote uniform tensile stress, σ_∞, is in the following form [Shih and Hutchinson, 1976]

$$\mathrm{J} = \mathrm{c}\left(n_p\right)\sigma_\infty \varepsilon_\infty a$$

$$= \left\{ 3.85\sqrt{n_p}\left(1 - \frac{1}{n_p}\right) + \frac{\pi}{n_p} \right\} \sigma_\infty \varepsilon_\infty a \qquad (2.22)$$

Other forms of $C(n_p)$ are proposed elsewhere as well.

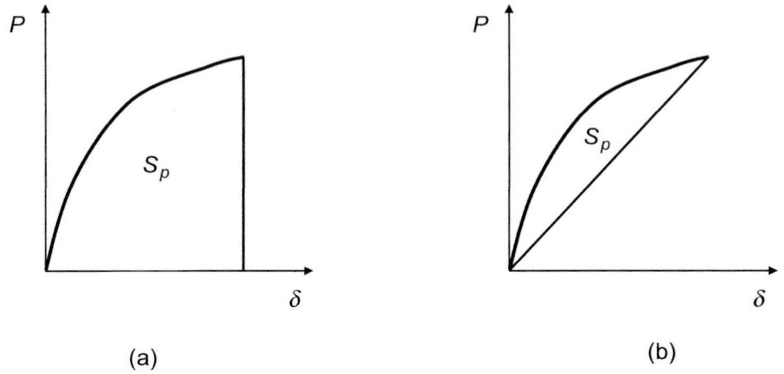

Figure 2.6 Area in load-displacement curve for evaluation of the J-integral. (a) Plate specimen under bending and (b) plate specimen with center crack under tension.

2.3 Fracture Toughness and Fatigue Crack Propagation

2.3.1 Fracture Toughness

Fracture toughness is a mechanical quantity that indicates the fracture resistance of a cracked material under increased load. The quantity identifies the starting point of crack propagation (or unstable one) and is usually expressed as a critical magnitude of fracture mechanics parameter. In a linear elastic body, including the small-scale yielding state, a crack begins to propagate when the applied linear elastic stress intensity factor (fracture driving force), K_{Iap}, reaches the material-characteristic critical value (fracture resistance of the material). The criterion differs for plane strain and plane stress conditions, indicating the existence of multiaxial stress (strain) effect. Usually, the one in plane strain is adopted as the representative value of fracture toughness in materials, denoted as K_{IC}, because it is lower compared to the one in plane stress.

Under large-scale yielding, where the plastic zone ahead of the crack tip is large, K_{IC} loses its physical meaning. In such cases, the J-integral becomes the governing parameter as the criterion, denoted as J_{IC} [Begley and Landes, 1972; Landes and Begley, 1972].

Considering the features of K and J as the energy release rate in a linear elastic body (Eq. (2.13)), J_{IC} is universally applicable not only in large-scale yielding but also in small-scale one.

2.3.2 Fatigue Crack Propagation

A crack stably propagates in cyclic loading, known as fatigue crack propagation. Numerous investigations have been conducted from the standpoint of fracture mechanics, and the propagation law has been reported.

Under small-scale yielding state, the crack propagation rate (crack extension per unit cycle), da/dN, is correlated with the difference between K at the maximum load, K_{max}, and at the minimum one, K_{min}, namely the stress intensity factor range, $\Delta K = K_{max} - K_{min}$. Figure 2.7 schematically illustrates the da/dN–ΔK relationship in fatigue at room temperature. Although there are three stages, the first and second ones are important in terms of engineering application. The crack propagation rate has power-law relation with ΔK at the steady-state crack propagation (second stage), while it does not propagate below a threshold stress intensity (first stage), ΔK_{th} [Paris, 1960].

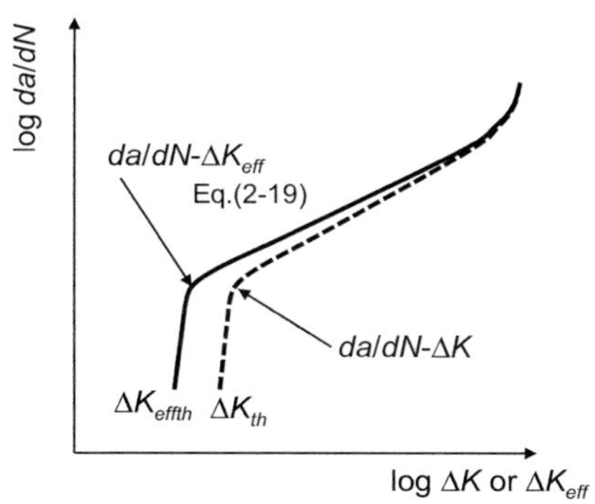

Figure 2.7 Crack propagation behavior in fatigue under the small-scale yielding conditions in fatigue, showing the relation between the crack propagation rate and the stress intensity factor ranges, ΔK or ΔK_{eff}.

Subsequent detailed research reveals that this relationship is affected by the closure of crack tip during unloading or compression. Then, the governing parameter is identified as the effective stress intensity factor range, $\Delta K_{eff} = K_{max} - K_{open}$, where K_{open} is K at the crack tip opening point [Elber, 1970]. The relation is schematically illustrated by the thick line in Fig. 2.7 and it is in the following form at the second stage.

$$\frac{da}{dN} = c_f \Delta K_{eff}^{m_{kf}} \quad c_f \text{ and } m_{kf} \text{ are material constants} \tag{2.23}$$

It should be noted that there is a threshold even in the da/dN–ΔK_{eff} relationship.

Under large-scale yielding, while ΔK_{eff} loses its effectiveness, the relation can be extended to

$$\frac{da}{dN} = \Delta J_f^{m_f} \tag{2.24}$$

using the cyclic J-integral (fatigue J-integral), $\Delta J_f = J_{max} - J_{open}$, where J_{max} and J_{open} are J at the maximum load and the crack tip opening point, respectively, (Fig. 2.8) [Dowling and Begley, 1976].

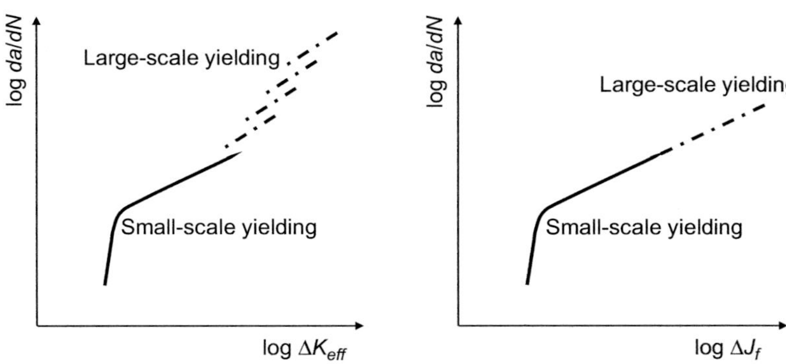

Figure 2.8 Crack propagation behavior in fatigue at room temperature under the small- and large-scale yielding conditions in fatigue, showing the da/dN–ΔK_{eff} relationship (left) and the da/dN–ΔJ_f relationship (right).

Applying the evaluation method of the J-integral, Eq. (2.19), the magnitude of ΔJ_f is simply obtained regardless of the constitutive (stress–strain) relation by

$$\Delta J_f = \frac{\Delta K^2}{E} + \frac{S_p}{h_t b} \tag{2.25}$$

where S_p is the area in the P–δ curve of the loading half cycle as depicted in Fig. 2.9 (center-cracked plate specimen). There is the threshold in the law, and the crack does not propagate below $\Delta J_{f\,th}$.

$$\frac{da}{dN} = 0 \quad \text{when} \quad \Delta J_f < \Delta J_{f\,th} \tag{2.26}$$

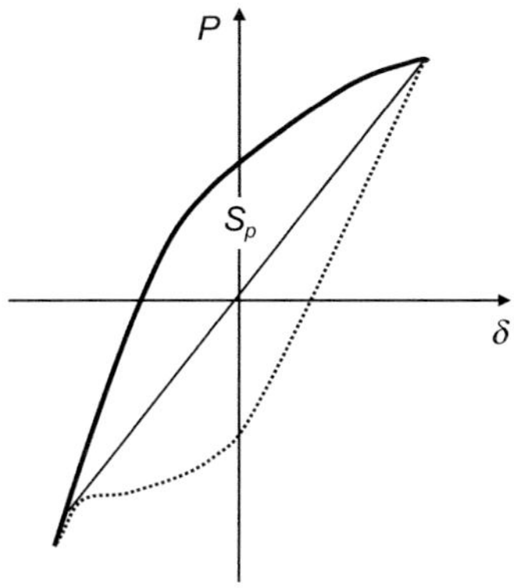

Figure 2.9 Load-displacement curve and area S_p for evaluation of the fatigue J-integral.

Appendices

2-A Path Independence [Rice, 1968a]

Figure 2A.1 illustrates two integral paths surrounding the crack tip from the lower crack face to the upper one, Γ_A and Γ_B, and their connection paths along the upper and lower crack faces, Γ_u and Γ_d, respectively. Since $n_1 = 0$ and $T_i = 0$ on Γ_u and Γ_d, the integral along the round path leads to the following relation:

$$\int_{\Gamma_A-\Gamma_u-\Gamma_B+\Gamma_d} \left(Wn_1 - T_i \frac{\partial u_i}{\partial x_1} \right) ds = \int_{\Gamma_A-\Gamma_B} \left(Wn_1 - T_i \frac{\partial u_i}{\partial x_1} \right) ds = J_A - J_B \tag{2A.1}$$

where J_A and J_B are the J-integrals evaluated on the paths, Γ_A and Γ_B, respectively. Substituting the relation,

$$T_\mathrm{i} = \sigma_\mathrm{ij} n_\mathrm{j} \tag{2A.2}$$

into the left-hand side of Eq. (2A.1) and applying the Green–Gauss theorem,

$$J_A - J_B = \int_D \left(\frac{\partial W}{\partial x_1} - \sigma_{ij} \frac{\partial^2 u_i}{\partial x_1 \partial x_j} - \frac{\partial \sigma_{ij}}{\partial x_j} \frac{\partial u_i}{\partial x_1} \right) dS \tag{2A.3}$$

is obtained. Substituting the relations

$$\frac{\partial \sigma_{ij}}{\partial x_j} = 0 \tag{2A.4}$$

and equilibrium equations

$$\frac{\partial W}{\partial x_1} = \frac{\partial W}{\partial \epsilon_{ij}} \frac{\partial \varepsilon_{ij}}{\partial x_1} = \sigma_{ij} \frac{\left(\frac{\partial^2 u_i}{\partial x_j \partial x_1} + \frac{\partial^2 u_j}{\partial x_i \partial x_1} \right)}{2} \tag{2A.5}$$

into Eq.(2A.3), we have

$$0 = J_A - J_B \tag{2A.6}$$

which means the J-integral is path-independent.

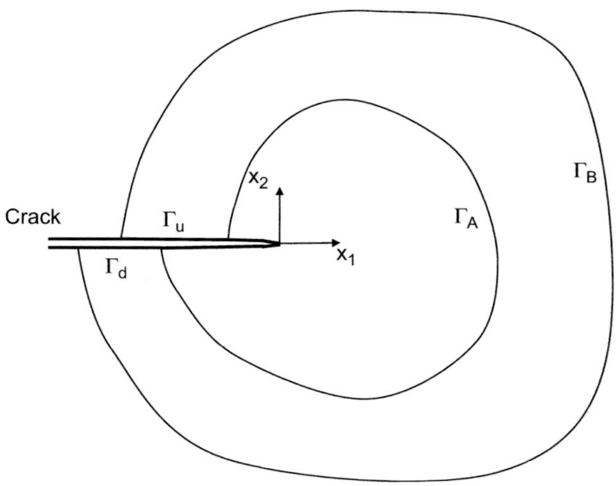

Figure 2A.1 Integral paths around a crack tip.

2-B Energy Release Rate [Rice, 1968b]

Consider a cracked body where the area D is surrounded by the path Γ and crack faces, as illustrated in Fig. 2B.1. The potential energy in

D, denoted as Π, is formulated as

$$\Pi = \int_D W dS - \int_\Gamma T_i u_i ds \tag{2B.1}$$

due to $T_y = 0$ on the crack faces. The change of Π under infinitesimal extension of the crack toward the x_1-direction is given by

$$\frac{d\Pi}{da} = \frac{d}{da}\int_D W dX dY - \int_\Gamma T_i \frac{du_i}{da} ds \tag{2B.2}$$

where X and Y are the coordinates fixed on the crack tip. Since $X = x_1 - a$, based on $\frac{\partial X}{\partial a} = -1$ and $\frac{\partial}{\partial X} = \frac{\partial}{\partial x_1}$ we have

$$\frac{d}{da} = \frac{\partial}{\partial a} - \frac{\partial}{\partial x_1} \tag{2B.3}$$

Then, Eq. (2B.2) becomes

$$\frac{d\Pi}{da} = \int_D \frac{\partial W}{\partial a} dX dY - \int_\Gamma W dY - \int_\Gamma T_i \frac{\partial u_i}{\partial a} ds + \int_\Gamma T_i \frac{\partial u_i}{\partial x} ds \tag{2B.4}$$

Since the first and third terms cancel each other due to the virtual work principle, we obtain

$$\frac{d\Pi}{da} = -\int_\Gamma W n_1 ds + \int_\Gamma T_i \frac{\partial u_i}{\partial x} ds \tag{2B.5}$$

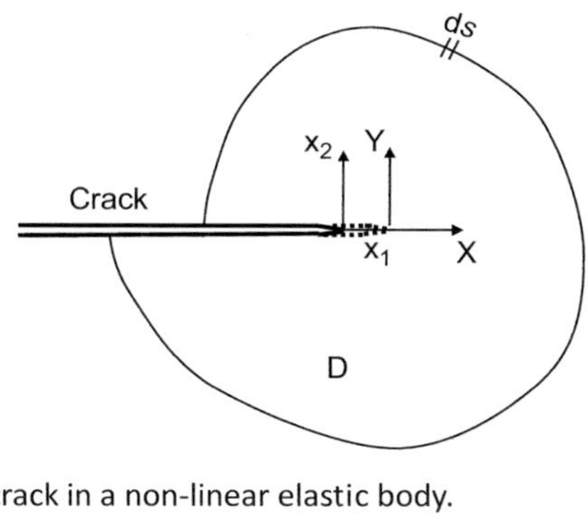

Figure 2B.1 A crack in a non-linear elastic body.

The right-hand side is the definition of the J-integral, Eq. (2.8). Thus, we have

$$\mathrm{J} = -\frac{d\Pi}{da} \tag{2B.6}$$

As shown in this section, we can define the J-integral as the energy release rate and can inversely derive Eq. (2.10).

2-C Stress and Strain Singular Fields Near Crack Tip and J-integral [Hutchinson, 1968; Rice and Rosengren, 1968]

Adopting a circle path with a radius of r_o for the integration as illustrated in Fig. 2C.1, the J-integral is in the form

$$\frac{J}{r_o} = \int_{\Gamma} (W\cos\theta - T_i \frac{\partial u_i}{\partial x}) d\theta \tag{2C.1}$$

Since the left-hand side is in the order of $1/r_o$, due to independence of J on r_o, average of the product of stress and strain along the circular path (the right-hand side) has a singularity of $1/r$, referring that r_o is arbitrary.

The following relation clearly satisfies the above requirement.

$$\sigma_{ij}\varepsilon_{ij} \to \frac{\text{function of } \theta}{r} \tag{2C.2}$$

Although it is unclear whether it is unique, it can serve as a guideline for investigating the stress and strain singular fields in a non-linear elastic body.

The above discussion holds true for a general elastic body. We now apply it to the following power-law non-linear elastic body.

$$\epsilon = B_p \sigma^{n_p} \qquad (2C.3) = (2.14)$$

Although the orthodox methodology is the analysis based on the stress function ϕ, it is too complex to solve analytically. Hence, as a dominant part of the function near the crack tip, we assume

$$\phi = K_\sigma r^s \breve{\phi}(\theta) \tag{2C.4}$$

where K_s is amplitude of $\breve{\phi}(\theta)$, s is an unknown constant. Then,

$$\sigma_{ij} = K_\sigma r^{s-2} \breve{\sigma}_{ij}(\theta) \tag{2C.5}$$

where $\breve{\sigma}_{ij}(\theta)$ is a function of only θ. Equations (2C.2) and (2C.3) lead to

$$s = \frac{2n_p + 1}{n_p + 1} \tag{2C.6}$$

The validity is checked by numerical analysis of eigenvalue equations derived from the exact stress function.

The singular fields of stress and strain near the crack tip are formulated as

$$\sigma_{ij} = K_\sigma r^{-1/(n_p+1)} \tilde{\sigma}_{ij}\left(\theta, n_p\right) \tag{2C.7}$$

$$\varepsilon_{ij} = K_\varepsilon r^{-n_p/(n_p+1)} \tilde{\varepsilon}_{ij}\left(\theta, n_p\right) \tag{2C.8}$$

Here K_σ and K_ε are the non-linear elastic stress intensity factor and non-linear elastic strain intensity factor, respectively.

Substituting Eqs. (2C.7) and (2C.8) into Eq. (2C.1), the relationships between J and K_σ, and J and K_ε are given by

$$J = B_p I_{n_p} K_\sigma^{\,n_p+1} \tag{2C.9}$$

$$J = \left(\frac{K_\varepsilon^{\,n_p+1}}{B_p}\right)^{\frac{1}{n_p}} I_{n_p} \tag{2C.10}$$

where I_{n_p} is a constant depending only on n_p. Then, Eqs. (2.15) and (2.16) are derived from (2C.7) to (2C.10)

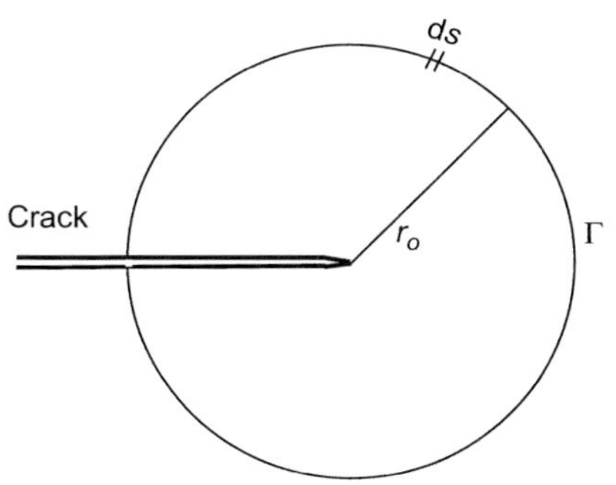

Figure 2C.1 Circular integral path for evaluation of the J-integral.

2-D Experimental Evaluation Method of J-Integral [Rice *et al.*, 1973]

Based on the definition of the energy release rate, the J-integral for a unit thickness is given by

$$J=\int_0^P\left(\frac{\partial\delta}{\partial a}\right)_P dP \quad \text{Under load-control} \tag{2D.1}$$

$$J=\int_0^\delta\left(\frac{\partial P}{\partial a}\right)_\delta d\delta \quad \text{Under displacement-control} \tag{2D.2}$$

For a plate specimen under bending, Eq. (2D.1) is rewritten as

$$J=\int_0^M\left(\frac{\partial\Theta}{\partial a}\right)_M dM \tag{2D.3}$$

where M is the applied moment and Θ is the angular displacement, respectively. Since deformation concentrates on the ligament for a deep crack (large a/wa), Θ becomes a function of the ligament length, b and M. As a dimension analysis leads to

$$\vartheta=g\left(\frac{M}{b^2}\right) \tag{2D.4}$$

we have

$$\left(\frac{\partial\vartheta}{\partial a}\right)_M=-\left(\frac{\partial\vartheta}{\partial b}\right)_M=\frac{2M}{b^2}\frac{\partial g}{\partial\left(M/b^2\right)}=\frac{2M}{b}\left(\frac{\partial\vartheta}{\partial M}\right)_b \tag{2D.5}$$

Substituting Eq. (2D.5) into Eq. (2D.4),

$$J=\frac{2}{b}\int_0^M M d\vartheta \tag{2D.6}$$

Considering the arm length, $M\vartheta\cong P\delta$, we have

$$J=\frac{2}{b}\int_0^P P d\delta \tag{2D.7}$$

When the plate thickness, h_b, is introduced in the above equation, this becomes Eq. (2.18)

For a plate specimen with a center crack under tension, Eq. (2.19) can be derived using a similar procedure.

References

Begley, J.A., and J.D. Landes, 1972, The J-integral as a fracture criterion, ASTM STP 514, pp. 1–20.

Dowling N.E., and J.A. Begley, 1976, Fatigue crack growth during gross plasticity and the J-integral, ASTM STP 590, pp. 82–103.

Elber, W., 1970, Fatigue crack closure under cyclic tension, *Engineering Fracture Mechanics*, 2, pp. 37–45.

Hutchinson, J.W., 1968, Singular behavior at the end of a tensile crack in a hardening material, *Journal of the Mechanics and Physics of Solids*, 16, pp. 13–31.

Irwin, G.R., 1956, Onset of fast crack propagation in high strength steel and aluminum alloys, *Naval Research Laboratory Report*, 4763.

Irwin, G.R., 1957, Analysis of stresses and strains near the end of a crack traversing a plate, *Journal of Applied Mechanics*, 24, pp. 361–364.

Landes, J.D., and J.A. Begley, 1972, The effect of specimen geometry on J_{IC}, ASTM STP 514, pp. 24–29.

Ohji, K., K. Ogura, and S. Kubo, 1978, Simplified evaluation method of J-integral under gross plasticity and application to creep cracking (in Japanese), *Transaction of Japanese Society of Mechanical Engineers*, 44, pp. 1831–1838. https://www.jstage.jst.go.jp/article/kikai1938/44/382/44_382_1831/_pdf/-char/ja

Paris, P.C., and J.A. Erdogan, 1960, A critical analysis of crack propagation laws, *Journal of Basic Engineering*, 85, pp. 528–534.

Rice, J.R., 1968a, A path independent integral and the approximate analysis of strain concentration by notches and cracks, *Journal of Applied Mechanics*, 35, pp. 379–386.

Rice, J.R. 1968b, Mathematical analysis in the mechanics of fracture, Chapter 3 In: *Fracture: An Advanced Treatise*, Vol. 2, Mathematical Fundamentals, ed. Liebowitz, Academic Press, N.Y., pp. 191–311.

Rice, J.R., and G.F. Rosengren, 1968, Plane strain deformation near a crack tip in a power-law hardening material, *Journal of the Mechanics and Physics in Solids*, 16, pp. 1–12.

Rice, J.R., P.C. Paris, and J.G. Merkle, 1973, Some further results of J-integral analysis and estimates, ASTM STP 536, pp. 231–245.

Shih, C.F., and J.W. Hutchinson, 1976, fully plastic solutions and large scale yielding estimation for plane stress problems, *Journal of Engineering Materials and Technology*, 98, pp. 289–295.

Williams, M.L., 1957, On the stress distribution at the base of a stationary crack, *Journal of Applied Mechanics*, 24, pp. 109–114.

Chapter 3

Creep Crack Propagation

3.1 Creep

3.1.1 Time-Dependent Deformation

Creep is known as a phenomenon that a material continuously deforms over time under a constant stress at a constant high temperature. Figure 3.1 schematically illustrates a typical creep curve, depicting the relationship between elongation and elapsed time, and the time-dependent deformation results in characteristic material strength at high temperatures. Instantaneous strain occurs at the initial loading, followed by characteristic time-dependent deformation. The creep curve can be classified into three regions:

(1) Transient creep: the strain rate decreases with time in the initial stage of creep.
(2) Steady-state creep: the strain rate remains constant in the middle stage of creep.
(3) Accelerated creep: the strain rate increases with time in the late stage of creep.

It has been discussed that accelerated creep is mainly caused by a reduction in cross-sectional area due to numerous small cracks and cavities in the material, and this may disappear under a true-stress-

High Temperature Fracture Mechanics
Takayuki Kitamura
Copyright © 2025 Jenny Stanford Publishing Pte. Ltd.
ISBN 978-981-5129-55-7 (Hardcover), 978-1-003-65328-8 (eBook)
www.jennystanford.com

controlled condition. After these stages, the material eventually ruptures.

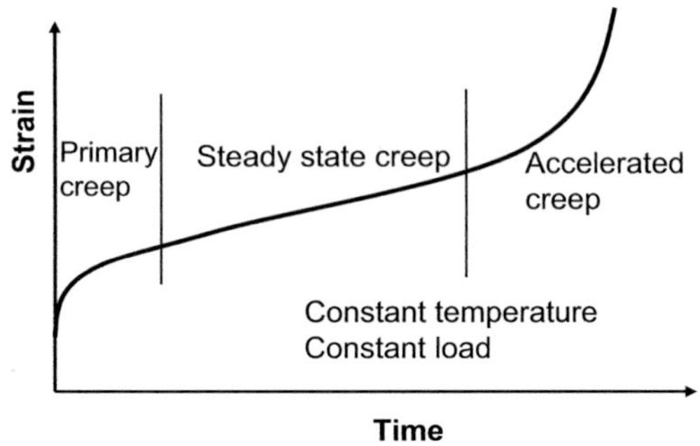

Figure 3.1 Schematic diagram of typical creep curve. Change in strain with time under constant a load at a constant high temperature.

In steady-state creep (2), there is a one-to-one correspondence between applied stress, σ, and strain rate, $\dot{\varepsilon}$. The relationship is often expressed by the following power-law, known as the Norton law:

$$\dot{\varepsilon} = B\sigma^n \tag{3.1}$$

or Normalized form

$$\frac{\dot{\varepsilon}}{\dot{\varepsilon}_o} = \left(\frac{\sigma}{\sigma_o}\right)^n \tag{3.1'}$$

where B and n are material constants. This chapter primarily utilizes the constitutive equation (the relationship between stress and strain rate) in steady-state creep due to its simplicity unless otherwise specified, though several other complex constitutive equations have been proposed.

The relationship, Eq. (3.1), is extended to include the temperature, T, as

$$\dot{\varepsilon} = B'\sigma^n \exp\left(-\frac{\Delta H}{RT}\right) \tag{3.2}$$

where R and ΔH are the gas constant and the activation energy, respectively. Since creep is intrinsically a thermally activated process, it exhibits a strong dependence on temperature, T.

In transient creep, the strain rate shows significant change over time, which makes the equation more complicated. Although various functional forms have been proposed for the constitutive equation, typical examples are

$$\dot{\varepsilon} = B\sigma^n t^m \quad \text{time hardening law} \tag{3.3}$$

$$\dot{\varepsilon} = B\sigma^n \varepsilon^m \quad \text{strain hardening law} \tag{3.4}$$

where m is a material constant.

Generally, steady-state creep is a state in which the work hardening due to strain accumulation is balanced by the microstructural recovery (softening). In accordance with this, the transient creep is understood as a transition from an initial state with low dislocation density (soft) to the balanced state (steady-state creep) where work hardening prevails due to the strain increase. The process can be formulated as follows:

$$\dot{\varepsilon} = B{\sigma_{\text{eff}}}^n \tag{3.5}$$

$$\sigma_{\text{eff}} = \sigma - \sigma_{\text{int}} \tag{3.6}$$

$$\dot{\sigma}_{\text{int}} = h_a \dot{\varepsilon} - r_e \tag{3.7}$$

where s_{ff} is the effective stress, s_{int} is the internal stress, h_a is the work hardening rate, and r_e r_e is the recovery rate, respectively.

A specific form of h_a and r_e is proposed as follows [Pugh and Robinson, 1978]:

$$h_a = c_h {\sigma_{\text{int}}}^{-n_a} \tag{3.8}$$

$$r_e = \mathrm{D}{\sigma_{\text{int}}}^{n-n_a} \tag{3.9}$$

where C_h, n_a, and D are material constants.

3.1.2 Hoff Analogy [Hoff, 1954]

In general, the analysis of stress and strain (rate) in a time-dependent non-linear body is complex and difficult. However, a convenient method called Hoff analogy [Hoff, 1954] is known in the case of steady-state creep, where a one-to-one relationship exists between stress and strain rate.

The analogy points out that by simply converting strain and displacement in analytical result of non-linear elastic body into strain rate and displacement rate, one can obtain stress and strain rate in the corresponding steady-state creep body (Appendix 2-A).

This allows us to directly apply the knowledge of non-linear elastic fracture mechanics to analysis in creep.

Since Eq. (3.1) has the form obtained by converting strain in Eq. (2.11) into strain rate, the Hoff analogy is applicable. The features of cracked bodies in non-linear elasticity explained in Chapter 2 can be applied to those in steady-state creep by simply replacing strain and displacement with strain rate and displacement rate, respectively. These provide us with a fundamental direction for fracture mechanics in creep cracking as a starting point.

3.2 Creep J-Integral

A kind of J-integral can be defined based on the Hoff analogy as follows

$$J^* = \int_{\Gamma} (W^* n_1 - T_i \frac{\partial \dot{u}_i}{\partial x}) ds \quad (3.10)$$

where

$$W^* = \int_0^{\dot{\varepsilon}} \sigma d\dot{\epsilon} \quad (3.11)$$

This concept is independently proposed as a fracture mechanics parameter for creep crack propagation in Refs. [Landes & Begley. 1976] and [Ohji *et al.*, 1976]. It is referred to as the C^* parameter, C^*, and the modified J-integral, J', respectively. Since then, various notations have been used for this parameter. To avoid confusion among readers, this book adopts the term and notation "creep J-integral" and "J^*," which is a unified terminology. Both groups initially reported their fundamental idea and results in company reports or proceedings in 1974. Additionally, the work of [Nikbin *et al.*, 1976; Harper and Ellison, 1977; Koterazawa and Mori, 1977] provided strong support in the early stages of investigation.

The Hoff analogy points out that the creep J-integral has characteristic features in a creep body similar to those of the J-integral in the corresponding non-linear elastic body. These features include:

(1) path independence of the integral and
(2) energy-rate release rate.

Although the physical meaning of (2) is not clear, it is useful in deriving experimental evaluation methods for the J^* value.

Assuming the Norton law, Eq. (3.1), the Hoff analogy leads to the following singular fields of stress and strain rate near the crack tip in creep, which correspond to the HRR singular fields [Hutchinson, 1968; Rice and Rosengren, 1968].

$$\sigma_{ij} = \left(\frac{J^*}{I_n B r} \right)^{\frac{1}{n+1}} \tilde{\sigma}_{ij}(\theta) \tag{3.12}$$

$$\dot{\varepsilon}_{ij} = \left(\frac{J^*}{I_n B r} \right)^{\frac{n}{n+1}} \tilde{\dot{\varepsilon}}_{ij}(\theta) \tag{3.13}$$

It should be noted that the intensity of the singular field is solely represented by the creep J-integral, J^*. Referring to the success of J-integral in cracking in a plastic (non-linear) body, the creep J-integral, J^* is a strong candidate for the dominant parameter in creep cracking.

Applying the analogy, the creep J-integral, J^*, in a creep plate with a small crack under a remote constant stress, σ_∞, and strain rate, $\dot{\varepsilon}_\infty$, can be approximately evaluated as [Shih and Hutchinson, 1976]

$$\begin{aligned} J^* &= \mathrm{c}(n)\sigma_\infty \dot{\varepsilon}_\infty a \\ &= \left\{ 3.85\sqrt{n}\left(1 - \tfrac{1}{n}\right) + \tfrac{\pi}{n} \right\} \sigma_\infty \dot{\varepsilon}_\infty a \end{aligned} \tag{3.14}$$

Similarly, the experimental evaluation equations can be readily derived. For example, the magnitude of the creep J-integral for a center cracked plate is evaluated by [Ohji *et al.*, 1978]

$$J^* = \frac{n-1}{n+1} \frac{P}{h_t b} \dot{V} \cong \sigma_{\mathrm{net}} \dot{V} \tag{3.15}$$

3.3 Law in Creep Crack Propagation

3.3.1 Crack Propagation Behavior

A plate specimen of carbon steel as shown in Fig. 3.2 with a center notch (CCP specimen: Center Cracked Plate) is subjected to a constant load at 673 K [Taira *et al.*, 1980]. To elucidate the superiority and inferiority of mechanical quantities, specimens that differ in size a factor of 10 are prepared.

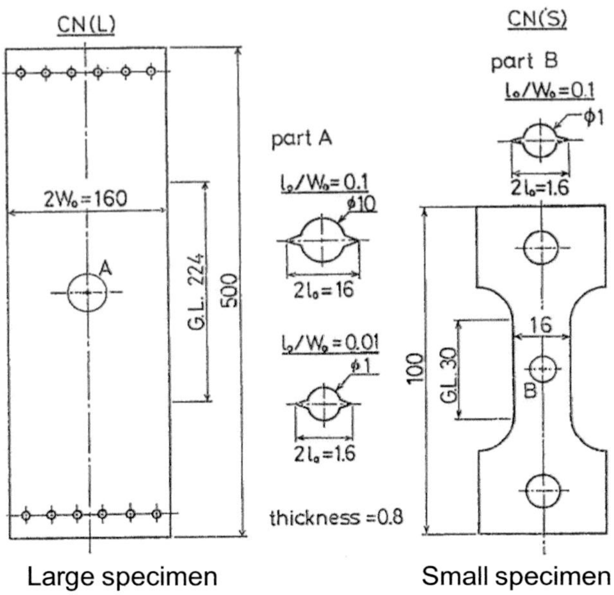

Figure 3.2 Center notched plate specimens for experiments of creep crack propagation [Taira *et al.*, 1980].

Figure 3.3(a) depicts the elongation of the large specimen in creep, showing similar behavior to the creep curve (Fig. 3.1). The arrow in the deformation curve indicates crack initiation from the notch root, revealing that the crack appears at a relatively early stage in creep. Since crack propagation accounts for most of the fracture life, careful consideration based on the fracture mechanics concept is crucial for engineering applications. In the later stage, deformation is significantly accelerated by the reduction of cross-section area due to crack propagation.

Figure 3.3(b) depicts the change in half crack length, including the notch depth. The crack propagation rate against time, da/dt, can be evaluated by the slope of the convex-shape curve. It is important to note that the half crack length is used for evaluation of da/dt because the center-cracked body has two tips to propagate. The figure clearly demonstrates that creep causes stable time-dependent crack propagation.

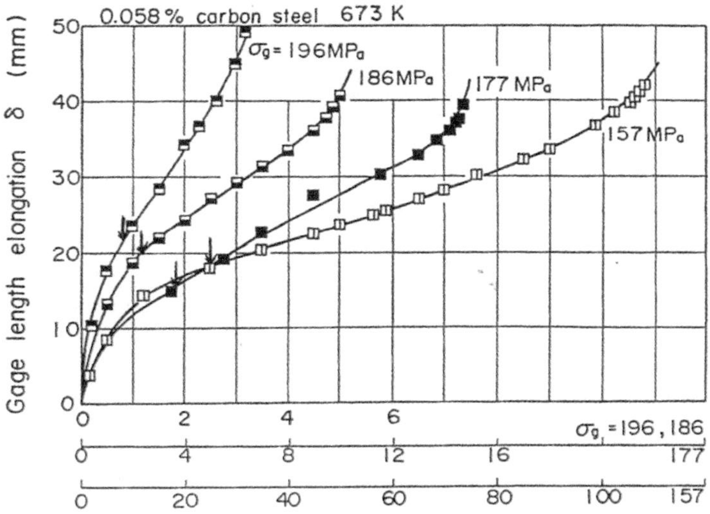

(a)

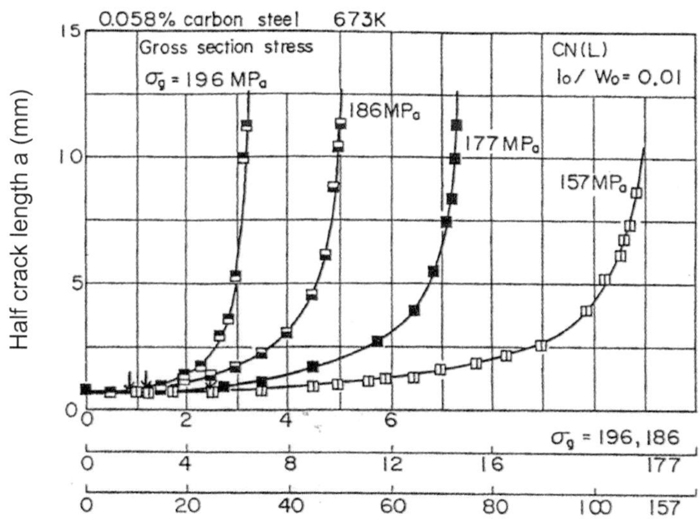

(b)

Figure 3.3 Crack propagation behavior in creep [Taira *et al.*, 1980]. (a) Time-dependent deformation at the loading point in creep crack propagation. (b) Time-dependent crack propagation in creep of plate specimen.

3.3.2 Fracture Mechanics Law

Figure 3.4 shows the relationships between the crack propagation rate, da/dt, and several mechanical parameters obtained from the experiments in the previous section [Taira *et al.*, 1980]. In addition to the da/dt–J^* relationship, the da/dt–K (linear elastic stress intensity factor), and da/dt–σ_{net} (net-section stress) ones are presented for comparison. Here, σ_{net} is the average stress on the cross-section of the ligament without stress concentration, indicating that the concentrated stress is fully relaxed. The large symbols represent the results for CCP specimens with a large plate width (2W = 160 mm) and the small symbols represent those with a small width (2W = 15 mm). Figure 3.4(a) and (b) clearly show that K and σ_{net} do not have a one-to-one relationship with da/dt. Figure 3.4(c) clearly indicates that J^* is the governing mechanical quantity in creep crack propagation, and the law is expressed as

$$da/dt = C_c J^{*m} \cong C_c J^* \tag{3.16}$$

where C_c and m are material constants. Interestingly, da/dt is approximately proportional to J^* ($m \cong 1$). It should be reminded that the fracture mechanics law can be applied to bodies regardless of dimensions of structure, making it applicable in various industrial fields, not only in power plants (larger size) but also in digital chips (smaller size), once the target material is tested in a laboratory.

We should be careful when determining the governing parameter through experiments. If we only examine the results for the small specimens (small symbols) in Fig. 3.4, da/dt shows a relatively good relationship with K and σ_{net} as well. Without theoretical knowledge, one might make a wrong judgement on the governing parameter. The experiments conducted on specimens of different sizes lead to the correct law, da/dt–J^* in the above experiments. It highlights that experiments conducted under conditions with the same near-field but different far-field are effective for determining the fracture mechanics law. This can be achieved by comparing the results of large and small specimens with large similarity ratios or by comparing the results of experiments on tensile and bending specimens.

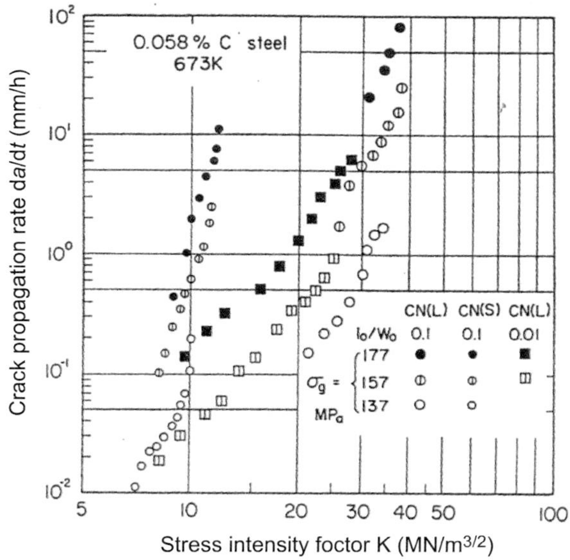
0.058 % C steel
673K
Crack propagation rate da/dt (mm/h)
Stress intensity foctor K (MN/m3/2)
CN(L) CN(S) CN(L)
lo/Wo 0.1 0.1 0.01
σg = 177 157 137 MPa
(a)

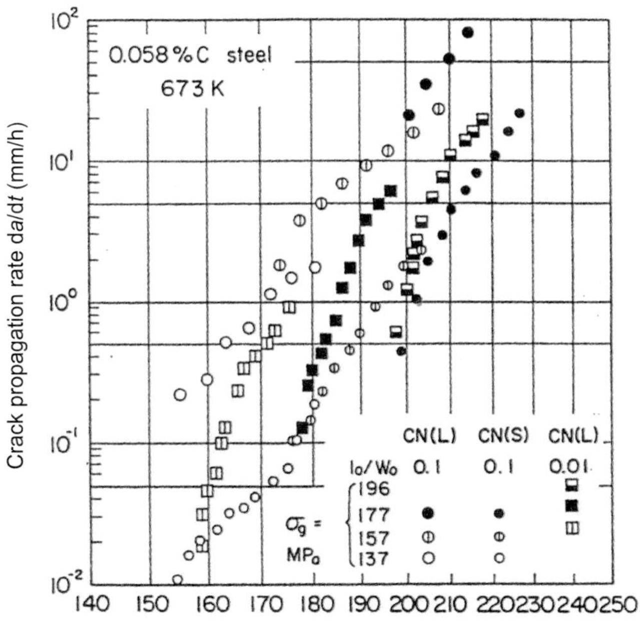
0.058%C steel
673 K
Crack propagation rate da/dt (mm/h)
Net-section stress σnet (MPa)
CN(L) CN(S) CN(L)
lo/Wo 0.1 0.1 0.01
σg = 196 177 157 137 MPa
(b)

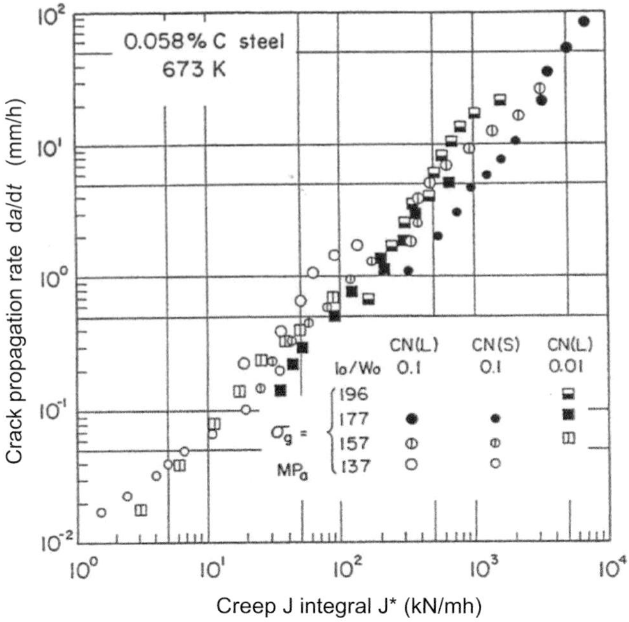

(c)

Figure 3.4 Examination on governing fracture mechanics parameter of the creep crack propagation rate [Taira *et al.*, 1980]. (a) Relationship between the crack propagation rate, d*a*/d*t*, and linear-elastic stress intensity factor, *K*. (b) Relationship between the crack propagation rate, d*a*/d*t*, and net-section stress, $\sigma_{net.}$ (c) Relationship between the crack propagation rate, d*a*/d*t*, and the creep J-integral, *J**.

3.3.3 Strain Rate Near Crack Tip

Specimens of carbon steel with a fine lattice micromesh attached around the crack are subjected to creep loading at 673 K, and both crack propagation and strain rate near the crack tip, $\dot{\varepsilon}_{c.t.}$, are simultaneously measured. Figure 3.5(a) [Kitamura, 1986] depicts the relationship between d*a*/d*t* and $\dot{\varepsilon}_{c.t.}$, indicating that crack propagation is clearly correlated with $\dot{\varepsilon}_{c.t.}$. Since *J** governs $\dot{\varepsilon}_{c.t.}$ as shown in Fig. 3.5(b), there is a strong correlation between d*a*/d*t* and *J** validating the concept of fracture mechanics.

If $m = 1$ in Eq. (3.16), it suggests that the strain energy rate near the crack tip is the critical factor for crack propagation. This also implies that the dependence of d*a*/d*t*–*J** relationship on the creep exponent n is weak. In a non-linear elastic (plastic) material, it is known that the J-integral has a proportional relationship with the crack tip opening displacement (CTOD). This suggests that the change in CTOD during an incremental time Δt has a good correlation with the crack extension Δa.

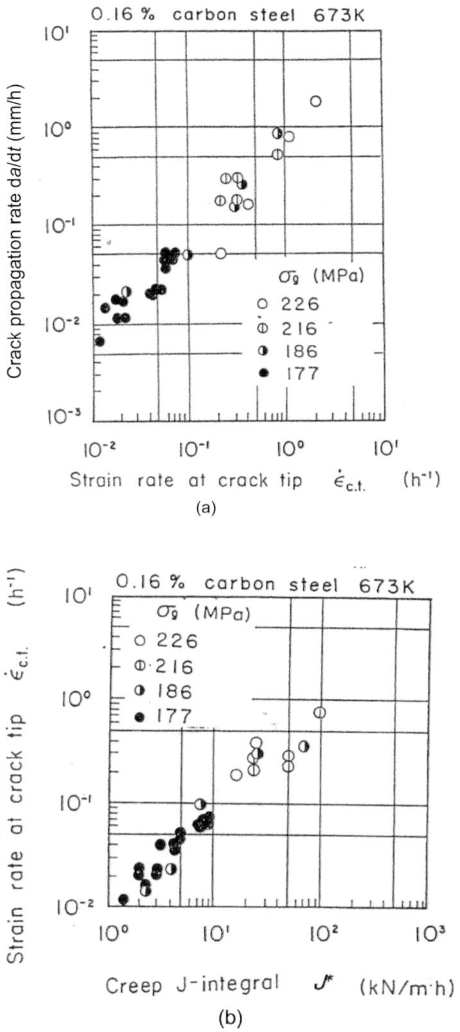

Figure 3.5 Strain rate near the crack tip experimentally observed by attached mesh in creep of carbon steel at 673 K [Kitamura, 1986]. (a) Relationship between the crack propagation rate and the strain rate near the crack tip. (b) Relationship between strain rate near the crack tip and the creep J-integral.

3.3.4 Influence Factors

There are many factors that influence creep crack propagation [Taira *et al.*, 1979] including the component size, which is discussed in the previous section (Fig. 3.4(c)). Although numerous research works have been conducted, this chapter explains some of the important influences below.

(1) Influence of temperature

Although creep strain rate is significantly influenced by temperature, Fig. 3.6 shows that the da/dt–J^* relation is independent of it [Taira *et al.*, 1979]. Equation (3.2) indicates that the temperature effect is reflected in the material constant B in the Norton law (Eq. (3.1)). Since $m \cong 1$ in Eq. (3.16) suggests that da/dt is closely correlated with the product of stress and strain rate near the crack tip, da/dt is approximately proportional to B. On the other hand, Eq. (3.10) indicates that J^* is proportional to B. The same dependence brings about that the da/dt–J^* relationship remains unaffected by temperature.

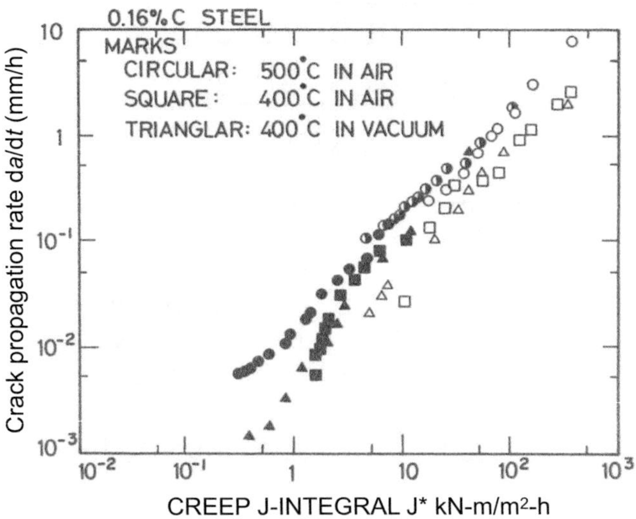

Figure 3.6 Influence of temperature on the da/dt–J^* relationship in creep of 0.16% carbon steel at 773 K and 673 K [Taira *et al.*, 1979].

It should be noted that the creep rupture mechanism (fracture resistance of material) may change significantly depending on temperature (for example, creep ductility). Under such conditions, the da/dt–J^* relationship may be affected by temperature.

(2) Influence of environment

The materials strength is susceptible to environmental effects such as oxidation at high temperatures. However, as shown in Fig. 3.7 [Taira *et al.*, 1979], the air environment has little influence on the crack propagation law, the da/dt–J^* relationship in creep of 304 stainless steel at 923 K under a vacuum of 10^2 Pa and normal air. This indicates that mechanical fracture (creep) prevails over chemically assisted fracture (oxidation) at the crack tip in mild environments even at high temperatures.

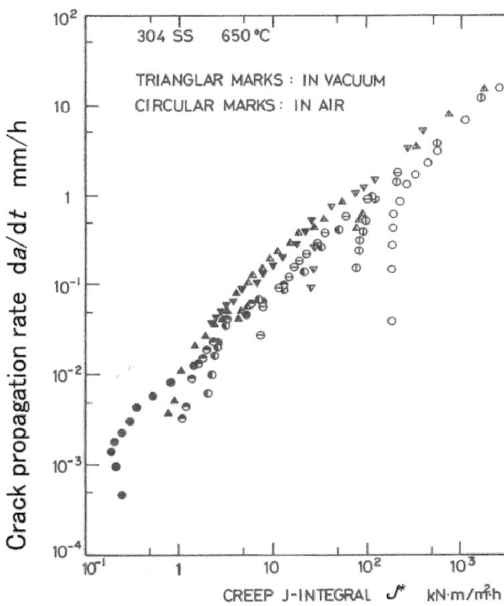

Figure 3.7 Influence of environment on the da/dt–J^* relationship in creep of 304 stainless steel [Taira *et al.*, 1979].

(3) Influence of pre-creep

Before introducing a starter crack, a smooth specimen of 316 stainless steel is subjected to creep until the strain reached 0.2 at 923 K [Taira *et al.*, 1979]. Considering that the fracture strain in creep is 0.22, the pre-damage introduced in the pre-creep specimen exceeds 90%. An experiment on crack propagation is conducted in creep at 923 K for the pre-creep specimen, comparing the result with those of non-damaged specimens. Fig. 3.8 shows the effect on the da/dt–J^* relationship, presenting that the pre-creep specimen clearly exhibits higher da/dt for the same J^* value. In other words, creep damage has a cumulative nature, and the fracture mechanics law depends on pre-creep.

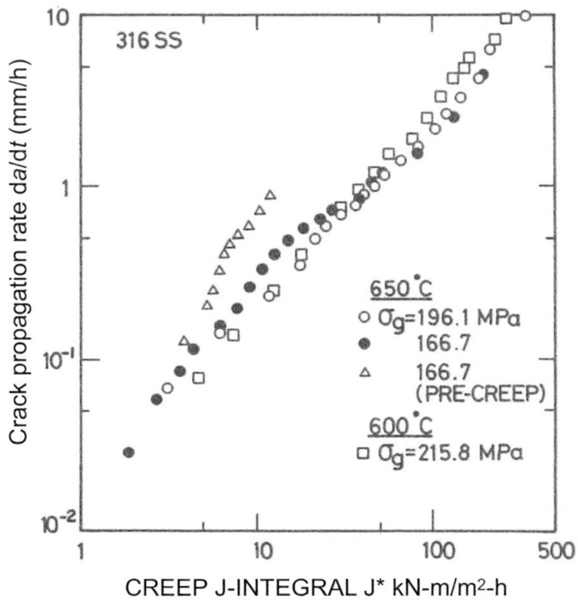

Figure 3.8 Influence of pre-creep strain on the da/dt–J^* relation in creep of 316 stainless steel at 923 K (650 °C) [Taira *et al.*, 1979].

Pre-creep materials can also be understood as those with a low creep fracture strain. There is a report that da/dt shows a universal relationship with J^*/ε_f, regardless of creep fracture strain ε_f [Nitta *et al.*, 1984]. However, this result does not quantitively agree with that.

(4) Influence of multiaxiality

Figure 3.9 shows the dependence of the da/dt–J^* relationship on plate thickness [Tabuchi *et al.*, 1997]. The crack propagation of thicker specimens is faster than that of thinner ones for the same J^* value suggesting that multiaxial stress (strain) strongly affects cracking (fracture at the crack tip). It should be reminded that this is similar to the result that the fracture toughness K_{Ic} of a thick specimen (under plane strain condition) is lower than that of a thinner one (under plane stress condition).

Mechanically, there is a method to analyze the influence of multiaxial stress using the deformation-constraint parameter due to hydraulic stress, Q. By applying this concept, the acceleration in creep cracking in thicker specimens can be explained by Q [Budden and Ainsworth, 1999].

It is known that creep fracture strain, e_f, decreases due to stress multiaxiality. From this perspective, it can be considered as a property closely related to (3) [Budden and Ainsworth, 1999].

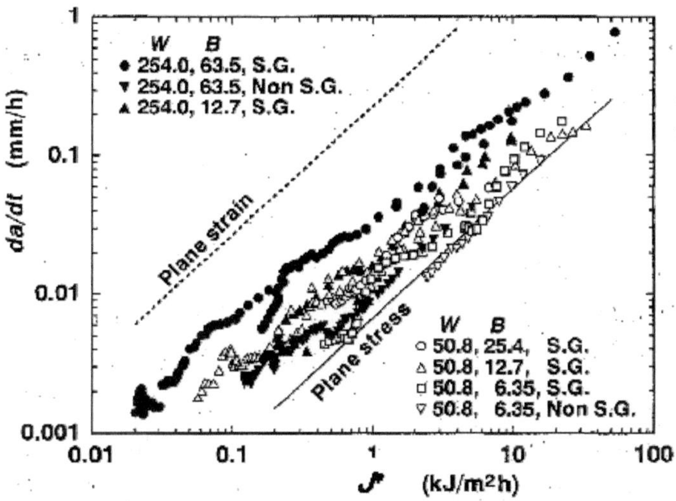

Figure 3.9 Influence of specimen size (plate-thickness) on the *da*/*dt*–*J** relationship in creep of 1Cr-Mo-V steel at 811 K [Tabuchi *et al.*, 1997].

3.4 d*a*/dt–*J** Relationships of Various Steels and Alloys

Figure 3.10 [Ohtani and Kitamura, 1988] shows a summary of crack propagation results obtained in the author's laboratory in creep of various materials under various temperatures. For the most part, the fracture mechanics law of Eq. (3.16) holds true over a wide range.

Additionally, differences between steels and alloys are relatively small.

In general, creep crack propagation is characterized by the following two factors:

(a) Intensity of stress and strain rate singular fields near the crack tip (fracture driving force).

(b) Failure conditions at the crack tip (fracture resistance).

The creep J-integral represents factor (a) as formulated in Eqs. (3.12) and (3.13), while C_c in Eq. (3.16) represents factor (b). Then, Fig. 3.10 suggests a relatively weak influence of factor (b), signifying that crack propagation is predominantly governed by the creep deformation behavior.

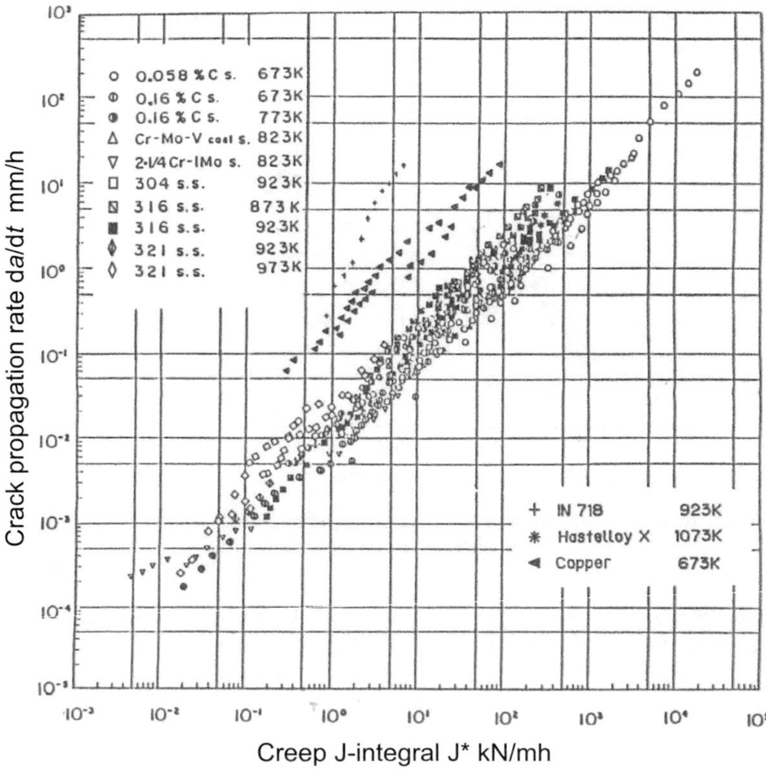

Figure 3.10 Creep crack propagation properties of various steels and alloys [Ohtani and Kitamura, 1988]. Note that data for IN 718 and copper requires verification.

Upon careful examination of Fig. 3.10, a trend emerges where da/dt tends to be lower for materials with a higher creep fracture strain for the same J^* value. Considering creep fracture strain, e_f, as a representative parameter of factor (b), there is a potential to develop a universal law from Eq. (3.16).

Appendix

3-A Analogy on Elasticity and Creep [Hoff, 1954]

The basic equations in steady-state creep are:

$$\sigma_{ij,j} = 0 \tag{3A.1}$$

$$\dot{\varepsilon}_{ij} = \frac{1}{2}\left(\dot{u}_{i,j} + \dot{u}_{j,i}\right) \tag{3A.2}$$

$$\dot{\varepsilon}_{ij} = F\left(\sigma_{ij}\right) \tag{3A.3}$$

On the other hand, those in a non-linear elastic body are:

$$\sigma_{ij,j} = 0 \tag{3A.4}$$

$$\varepsilon_{ij} = \frac{1}{2}\left(u_{i,j} + u_{j,i}\right) \tag{3A.5}$$

$$\varepsilon_{ij} = F\left(\sigma_{ij}\right) \tag{3A.6}$$

By converting the notations of σ_{ij} and u_i in Eqs. (3A.4) to (3A.6) into $\dot{\varepsilon}_{ij}$ and $\dot{u}_i$, we obtain the equations of steady-state creep body, (3A.1) to (3A.3). Since the only difference lies in the notation, the analysis in a non-linear elastic body can be applied to the corresponding creep body simply by the conversion.

Of course, this analogy is applicable not only to cracked bodies but also general structures in steady-state creep.

References

Budden, P.J., and R.A. Ainsworth, 1999, The effect of constraint on creep fracture assessments, *International Journal of Fracture*, 97, pp. 237–247.

Harper, M.P., and E.G. Ellison, 1977, The use of C* parameter in predicting creep crack propagation, *Journal of Strain Analysis for Engineering Design*, 12, pp. 167–179.

Hoff, N.J., 1954, Approximate analysis of structures in the presence of moderately large creep deformations, *Quarterly of Applied Mathematics*, 12, pp. 49–55.

Hutchinson, J.W., 1968, Singular behavior at the end of a tensile crack in a hardening material, *Journal of the Mechanics and Physics of Solids*, 16, pp. 13–31.

Kitamura, T., 1986, *Fracture Mechanics Investigation on Crack Propagation in High Temperature Fatigue*, Doctoral Theses.

Koterazawa, R., and T. Mori, 1977, Applicability of fracture mechanics parameters to crack propagation under creep condition, *Journal of Engineering Materials and Technology*, 99, pp. 298–305.

Landes, J.D., and J.A. Begley, 1976, A fracture mechanics approach to creep crack growth, ASTM STP 590, pp. 128–148.

Nikbin, K.M., G.A. Webster, and C.E. Turner, 1976, Relevance of nonlinear fracture mechanics to creep crack growth, ASTM STP 601, pp. 47–62.

Nitta, A., T. Kitamura, and K. Kuwabara, 1984, Creep crack propagation behavior of Type 321 stainless steel after long service in boiler, *Journal of The Society of Materials Science Japan* (in Japanese), 33, pp. 185–190. https://www.jstage.jst.go.jp/article/jsms1963/33/365/33_365_185/_pdf/-char/ja

Ohji, K., K. Ogura, and S. Kubo, 1976, Analysis on creep crack growth in mainly Mode III (in Japanese), *Transaction of Japanese Society of Mechanical Engineers*, 42, pp. 350–358. https://www.jstage.jst.go.jp/article/kikai1938/42/354/42_354_350/_pdf/-char/ja

Ohji, K., K. Ogura, and S. Kubo, 1978, Simplified evaluation method of J-integral under gross plasticity and application to creep cracking (in Japanese), *Transaction of Japanese Society of Mechanical Engineers*, 44, pp. 1831–1838. https://www.jstage.jst.go.jp/article/kikai1938/44/382/44_382_1831/_pdf/-char/ja

Ohtani, R., and T. Kitamura, 1988, Characterization of high temperature strength of metals based on the mechanics of crack propagation, In *High Temperature Creep-Fatigue*, R. Ohtani, M. Ohnami and T. Inoue (eds.), Elsevier Applied Science, London, 65–90.

Pugh, C.E., and D.N. Robinson, 1978, Some trends in constitutive equation model development for high-temperature behavior of fast-reactor structural alloys, *Nuclear Engineering and Design*, 48, pp. 269–276.

Rice, J.R., and G.F. Rosengren, 1968, Plane strain deformation near a crack tip in a power-law hardening material, *Journal of the Mechanics and Physics in Solids*, 16, pp. 1–12.

Shih, C.F., and J.W. Hutchinson, 1976, Fully plastic solutions and large scale yielding estimation for plane stress problems, *Journal of Engineering Materials and Technology*, 98, pp. 289–295.

Tabuchi, M., K. Kubo, and K. Yagi, 1997, Effect of specimen thickness on creep crack growth properties of 1Cr-Mo-V steel, *Journal of The Society of Materials Science Japan* (in Japanese), 46, pp. 53–57. https://www.jstage.jst.go.jp/article/jsms1963/46/1/46_1_53/_pdf/-char/ja

Taira, S., R. Ohtani, and T. Kitamura 1979, Application of J-integral to high-temperature crack propagation, Part I: Creep crack propagation, *Journal of Engineering Materials and Technology*, 101, pp. 154–161.

Taira, S., R. Ohtani, R. Shimizu, T. Kitamura, and T. Kashiwagi, 1980, Investigation on governing parameter of creep crack propagation using cracked specimens of wide-thin-steel (in Japanese), *Transaction of Japanese Society of Mechanical Engineers*, 46, pp. 468–476. https://www.jstage.jst.go.jp/article/kikaia1979/46/405/46_405_468/_pdf/-char/ja

Chapter 4

Influence of Stress Change on Creep Crack Propagation

4.1 Influence of Mild Stress Change

4.1.1 Crack Propagation in Dynamic Creep

Creep takes place under various stress conditions. Dynamic creep is the one under sustained stress superimposing vibration as depicted Fig. 4.1(b), and creep under constant stress is often called static creep (Fig. 4.1(a)) to distinguish from it. Here, the stress ratio, R, is defined as $R = {}^{\sigma_{\min}}/_{\sigma_{\max}}$ where $\sigma_{\max}$ and $\sigma_{\min}$ are the maximum and minimum stresses. The amplitude increases as R decreases, and the condition approaches fatigue loading. It is known that the rapture life of a smooth specimen without a crack or a notch in dynamic creep can be correlated well with that in static creep with integration of the small stress fluctuation, suggesting that the influence of vibration with small amplitude is not significant in creep behavior. Then, as a first step to examine the influence of stress change on creep crack propagation, the characteristics in dynamic creep are investigated.

High Temperature Fracture Mechanics
Takayuki Kitamura
Copyright © 2025 Jenny Stanford Publishing Pte. Ltd.
ISBN 978-981-5129-55-7 (Hardcover), 978-1-003-65328-8 (eBook)
www.jennystanford.com

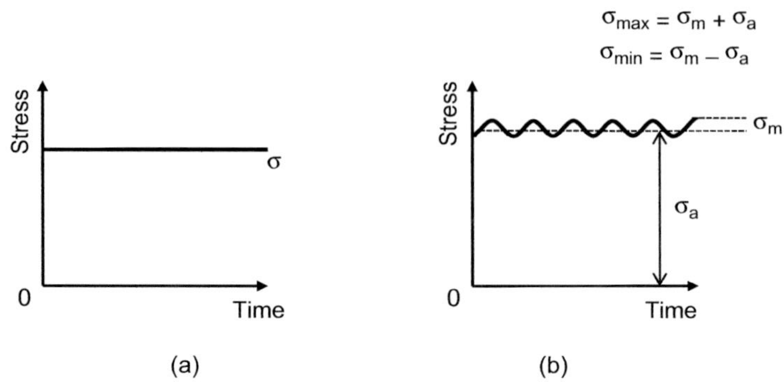

Figure 4.1 Applied stress of static and dynamic creep. (a) Static creep. (b) Dynamic creep.

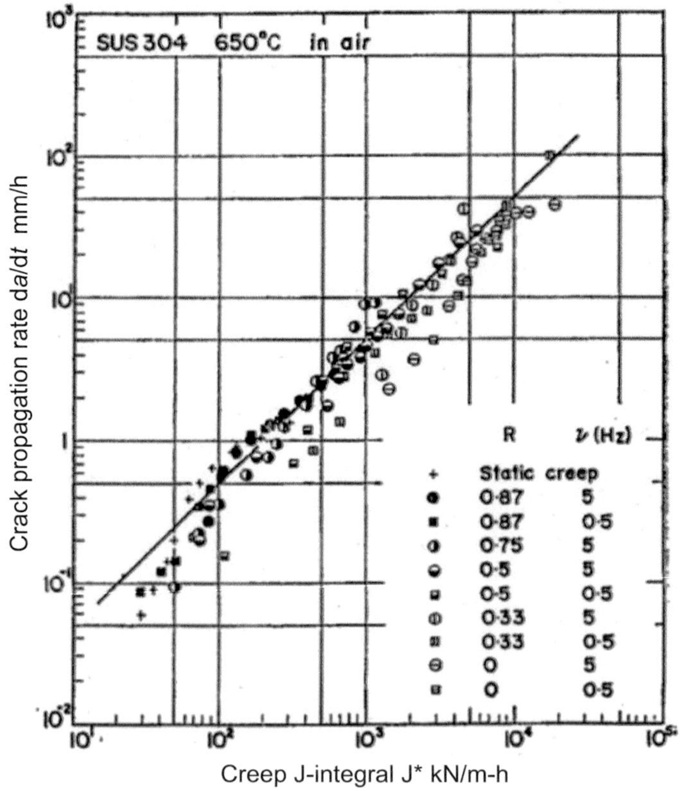

Figure 4.2 Applicability of creep J-integral to crack propagation in dynamic creep of 304 stainless steel at 923 K [Ohtani *et al.*, 1982].

Figure 4.2 shows crack propagation behavior in dynamic creep compared with that in static creep [Ohtani *et al.*, 1982], where the crack propagation rate, da/dt, is evaluated as the average in several cycles. da/dt is correlated well with the creep J-integral, J^*, in dynamic creep with $R > 0.5$, and the da/dt–J^* relationship coincides with that in static creep. Thus, the fracture mechanics law in creep holds even in creep with mild stress change of vibration. From the stress condition alone, cyclic loading with large stress amplitude should be classified in fatigue rather than dynamic creep. However, the results for stress ratio $R < 0.5$ in Fig. 4.2 show that the da/dt–J^* relationship is still maintained even when R = 0, indicating creep-dominated crack propagation.

For the experiments conducted under $R < 0$ (tension/compression conditions), although those results are not shown in the figure, the da/dt–J^* relationship does not hold. This suggests that there is crack propagation caused by a different mechanism from creep, namely fatigue crack propagation which is explained in Chapter 5. The boundary between creep crack propagation (time-dependent) and fatigue one (cycle-dependent) is further explained in Section 7.3 (Chapter 7).

4.1.2 Crack Propagation under Slow Decline of Load

Since the J-integral is defined for a non-linear elastic body, it is effective only under increasing (or constant) load conditions for a plastic body. It cannot be used to analyze crack propagation under declining load conditions. In the case of fatigue, as explained in Section 2.3.2 (Chapter 2), the J-integral is applied only to the increasing-load half cycle.

Crack propagation experiments in static creep with a constant load causes the net-section stress to increase as the crack propagates, resulting in increase of the creep J-integral, J^*. On the other hand, creep damage progresses despite a decrease of load at high temperatures. For example, when the applied strain is kept constant, stress decreases as shown in Fig. 4.3, namely stress relaxation, and shortens the failure life of materials. While the total strain remains constant, creep strain occurs over time due to the applied stress while elastoplastic strain gradually reduces. This suggests that creep strain near the crack tip increases even under declining stress conditions and causes creep crack propagation.

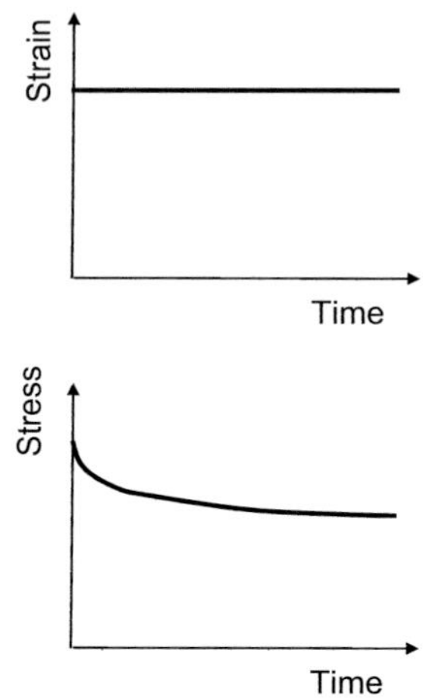

Figure 4.3 Stress drop (relaxation) under strain hold at a high temperature.

Importantly, the creep J-integral is derived using the Hoff analogy to the J-integral in a non-linear elastic body. The J-integral is applicable even under declining load conditions if the stress–strain relationship is reversible without hysteresis, unlike plastic body. As the creep J-integral is evaluated at any instant, it meets the requirement of reversibility in the stress–strain rate relationship when the stress changing rate is slow.
The relationship can be expressed as

$$\dot{\varepsilon} = B\sigma(t)^n \tag{4.1}$$

Since the Norton law governs the creep-cracked body at any given instance, the creep J-integral remains effective even under declining load conditions. Furthermore, the evaluation method explained in Chapter 3 remains valid at any instance [Ohtani *et al.*, 1988].

Figure 4.4 shows the relationship between da/dt and J^* under slow declining load [Ohtani *et al.*, 1988]. The crack propagation rate is governed by the creep J-integral, and the relationship coincides well with that in static creep. It should be noted that it holds for creep with repeating (cyclic) declining load as well.

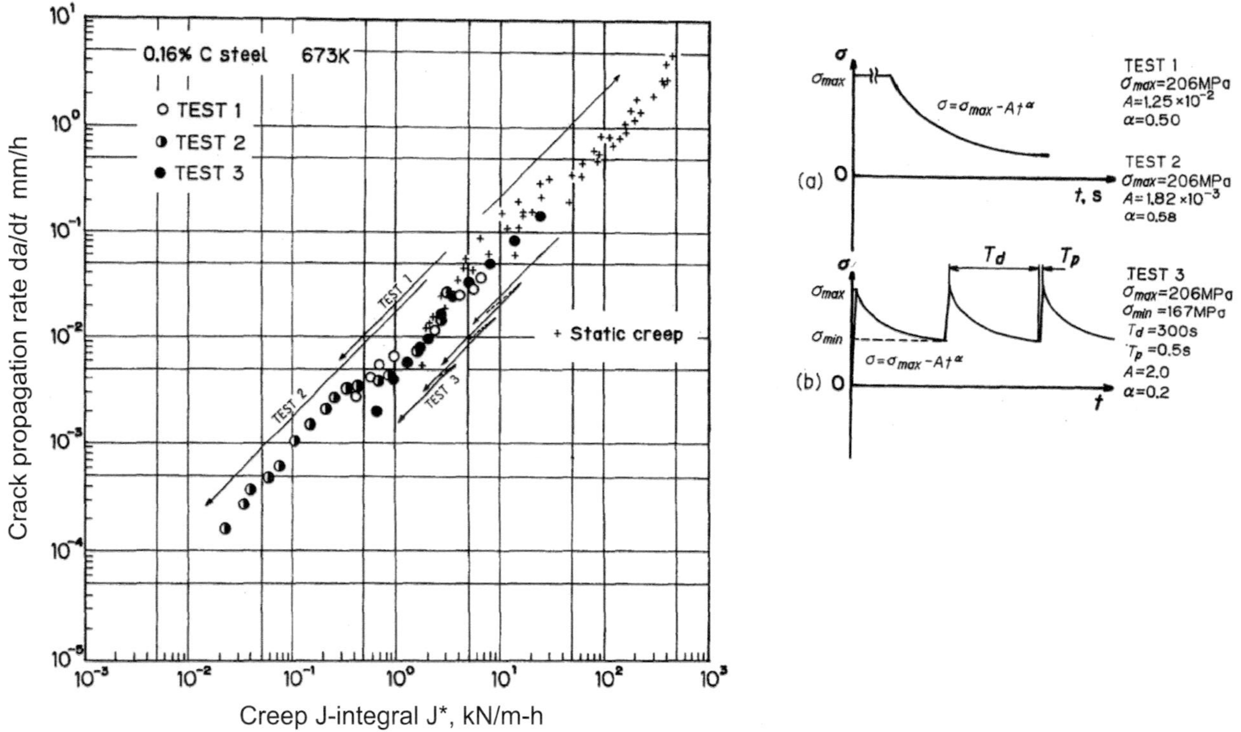

Figure 4.4 Relationship between the crack propagation rate and the creep J-integral under declining stress creep of carbon steel at 673 K [Ohtani *et al.*, 1988].

4.2 Influence of Step Load

It is reported [Taira *et al.*, 1979] that crack propagation a sudden (step-like) load change behaves as schematically illustrated in Fig. 4.5. In the case of a step-up load (left in Fig. 4.5(a)), the crack propagation rate, da/dt, overshoots immediately after the

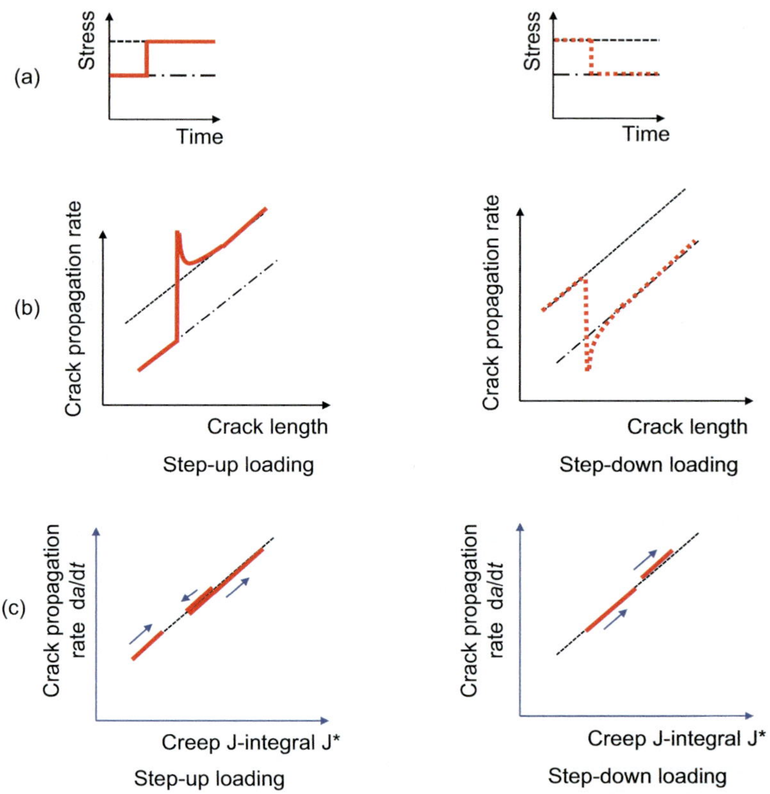

Figure 4.5 Schematic diagrams showing creep crack propagation under two-step loadings. (a) Step-up and step-down loadings in creep. (b) Transient crack propagation behavior. (c) Relationships between the crack propagation rate and the creep J-integral.

stress change, and then gradually converges to da/dt in constant high load. It should be noted that there is clear transient crack propagation after the step load for a certain period. The strain rate in creep is understood as the competition of strain-hardening and microstructural recovery (softening), and the balanced condition of steady-state creep depends on the stress level, pointing out that they differ between high and low loads. Thus, the step-up load induces a transition from equilibrium at low to that at high stress level, resulting in appearance of transient creep after the load change. Of course, it eventually converses to the new equilibrium state after a certain period.

Similarly, in the case of a step-down load (right in Fig. 4.5(b)), da/dt also overshoots immediately after the stress change, and then gradually converges to da/dt in constant low load. The transition in the step-down load can be explained similarly to that in the step-up load.

It is reported [Taira *et al.*, 1979] that the relationship between da/dt and J^* in step loads and corresponding constant loads is as schematically illustrated in Fig. 4.5(c). Note that this creep J-integral takes into account the acceleration/deceleration of transient creep strain rate, which will be explained in Section 4.3. da/dt is proportional to J^* in both of step loads including the transient period, and the da/dt–J^* relationship coincides well with that in constant loads. The fracture mechanics law in creep, Eq. (3.16), remains effective not only in constant load but also in step–stress change.

4.3 Factors in Transient Stress Singularity

For discussion on creep crack propagation under transient conditions after load change, there are two influence factors:

(1) Material factor: This is caused by the transition of stress–strain rate relationship (deformation property) due to the transient change of microstructure. The transition occurs not only in a cracked body but also in a creep body without a crack or notch.

(2) Mechanical factor: While the elastoplastic strain is dependent only on stress, the creep strain depends not only on stress but also on time elapsed. While the former dominates the entire body immediately after loading, the latter prevails as time elapses. When there is stress distribution in a body, the elastoplastic stress field dominates initially but is gradually superseded by the creep one in the long term, reaching creep-dominated field at last.

Crack propagation due to each factor is explained in the following Sections 4.4 and 4.5, respectively.

4.4 Singularity in Transient Creep: Material Factor

4.4.1 Creep J-Integral

In transient creep, it is not easy to analyze the singular fields of stress and strain rate near the crack tip because the exact constitutive equation is complex and not clearly formulated. The simplest one is the time-hardening law, Eq. (3.3), which indicates that B in Eq. (3.1) is time-dependent, $B(t)$.

$$\dot{\varepsilon} = B\sigma^n t^m = B(t)\sigma^n \tag{4.2}$$

Then, the creep J-integral, J^*, can be defined at each instance, capturing all the features. It should be noted that the magnitude of J^* depends on time, $J^*(t)$, due to the time-dependence of $B(t)$. Recalling the discussion of temperature influence on the $\mathrm{d}a/\mathrm{d}t$–J^* relationship, its relation is expected to be independent of the transition.

In the case of strain-hardening law, Eq. (3.4), assuming proportional loading, a similar conclusion is derived [Ohji *et al.*, 1980; Riedel, 1981].

Detailed analysis is conducted on the transition of the singular stress by a step loading based on more precise constitutive equations, Eqs. (3.5) to (3.9), accounting for the transition due to work hardening and recovery [Kubo, 1983]. According to the analysis, as illustrated in Fig. 4.6, steady-state creep usually becomes predominant very near the crack tip immediately after loading, while transient creep governs the distant region. As time elapses, the steady-state creep region near the crack tip expands, and the transition is completed when it covers most of the cracked body. Of course, afterward, the features of steady-state creep are fully applicable.

In the transient condition, the singular fields of stress and strain rate near the crack tip obey Eqs. (3.12) and (3.13) because the area is governed by steady-state creep. This implies that the intensity is given by J^*. However, it should be noted that the magnitude changes over time, $J^*(t)$, because the steady-state creep area is surrounded by transient creep. Moreover, it is reported that $J^*(t)$ can be estimated by the integration path apart from the crack tip [Kubo, 1983]. In the case of step-up load, considering the transition (the early stage of creep curve of Fig. 3.1), the temporal change of J^* is schematically illustrated in Fig. 4.7. The step-down load is discussed similarly in the literature [Kubo, 1983].

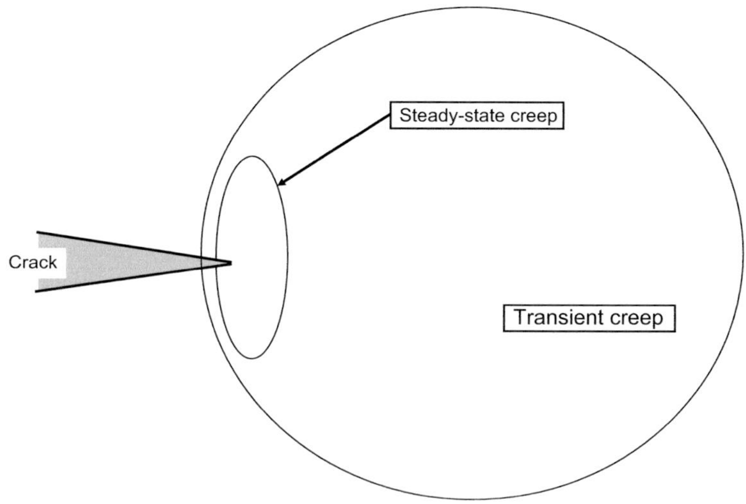

Figure 4.6 Development of steady-state creep region near the crack tip under transient creep [Kubo, 1983].

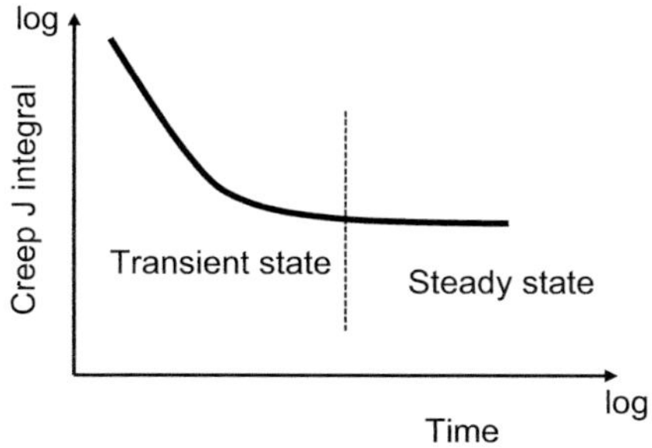

Figure 4.7 Schematic diagram of change in the creep J-integral with loading time under transient creep.

4.4.2 C_t Parameter

The constitutive equation in transient creep is highly dependent on the loading history, making the analysis of singular fields near the crack tip difficult. Considering this complexity, another extension from the creep J-integral in steady-state creep is proposed, denoted as C_t, [Saxena, 1986]. It is defined as the energy-rate release rate formulated by

$$C_t = -\frac{d\Pi^*}{da} \tag{4.3}$$

where Π^* is the potential energy rate defined as a parameter where the displacement and strain in the potential energy of elastic body are converted into their respective rates, following the Hoff analogy. Although its physical interpretation is unclear, C_t can be easily evaluated using the relationship between load and displacement rate, regardless of the constitutive equation. As explained in Section 3.2 (Chapter 3)

$$C_t = J^* \tag{4.4}$$

in steady-state creep. Recalling the features explained in previous section (Section 4.4.1). C_t closely resembles J^* even in transient creep.

Considering these overall features, the "creep J-integral and J^*" are used in a unified form throughout this book to avoid confusion.

4.4.3 Experimental Verification

The effectiveness of the above analysis is confirmed by the experimental results under step loads in Section 4.2. When the load suddenly changes, a characteristic transient creep is reflected in the change in the creep J-integral as well as the crack propagation rate, resulting in the same da/dt–J^* relationship as that in static creep. Note that this indicates the effectiveness not only J^* but also C_t (and $C(t)$ introduced in the next section).

4.5 Singularity in Small-Scale Creep: Mechanical Factor

4.5.1 Transition due to Expansion of Creep Area

Consider a linear elastic creep body subjected to a step load with the following constitutive equation:

$$\dot{\varepsilon} = \frac{\dot{\sigma}}{E} + B\sigma^n \tag{4.5}$$

where $\dot{\sigma}$ is stress rate. At the loading instance, since there exists no creep strain, the strain in the entire cracked body must be linear elastic, and the stress singular field near the crack tip is represented by the linear elastic intensity factor, K, as formulated in Eq. (2.2). While creep strain occurs over time, it preferentially appears near the crack tip due to higher stress. Shortly after loading, the creep-dominant region near the crack tip is surrounded by the linear-elastic area. As time progresses, this region expands as illustrated in Fig. 4.8(b), and it reaches the steady-state creep condition after the creep-dominated region covers the entire body. This transition from a linear elastic state to a fully creep state is due to interaction between elasticity and creep.

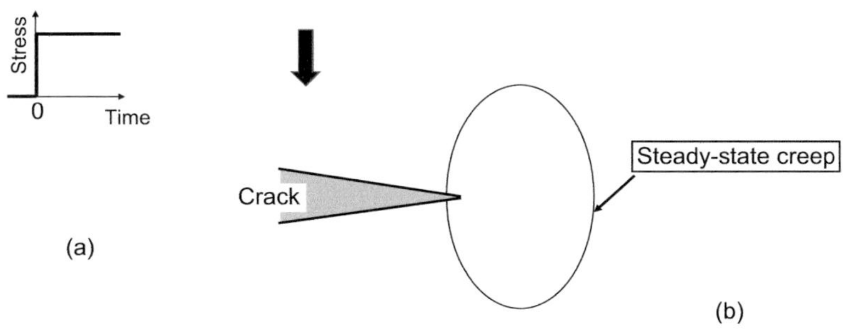

Figure 4.8 Step loading and transition from small-scale creep to large-scale creep. (a) Step loading. (b) Growth of creep-dominant area from the crack tip.

The state where the creep-dominated region is confined near the crack tip is called small-scale creep, while the state where the creep-dominated region covers a large area is called extensive creep or large-scale creep [Ohji *et al.*, 1980; Riedel and Rice, 1980]. These terms are named with reference to small-scale yielding of an elastoplastic cracked body under low load and to large-scale yielding under high load, respectively.

The Hoff analogy cannot be applied because the transient state includes not only steady-state creep but also linear elastic deformation. The creep J-integral explained in Section 3.2 needs to be fully reconsidered. Although the creep J-integral can be formally evaluated using Eq. (3.10), its value depends on the integral path.

Figure 4.9 [Ohji *et al.*, 1980] shows the change in the creep J-integral for the integral paths, obtained from finite element analysis for a center cracked plate of an elastic-creep body. Not only does the J^* value depend on the paths, but it also changes over time. The differences due to the integration paths become smaller over time, and the creep J-integral becomes constant when they converge. The converged point implies the end of transition.

On the other hand, the difference in the J^* magnitude is not large in the integral paths close to the crack tip even in the short-time after the step loading because the path independence of J^* holds for the paths within the creep-dominated region where the Norton law prevails and the Hoff analogy is satisfied. Limited to the crack tip region where creep is dominant, the fields of stress and strain rate near the crack tip are expressed by the HRR singularity, Eqs. (3.12) and (3.13). It should be noted that the J^* magnitude must be evaluated from an integral path within the creep-dominated region. As explained in Appendix 4-A, J^* during the transition is given by [Ohji *et al.*, 1980; Riedel and Rice, 1980]:

$$J^* = \frac{J_{el}}{(n+1)t} \tag{4.6}$$

This clearly indicates that the intensity decreases as the transition proceeds. In an elastic-plastic cracked body, the stress intensity near the crack tip is greater in large-scale yielding than in small-scale yielding. However, it should be noted that the elastic-creep body has the opposite feature; the stress intensity near the crack tip is greater in small-scale creep than in large-scale creep.

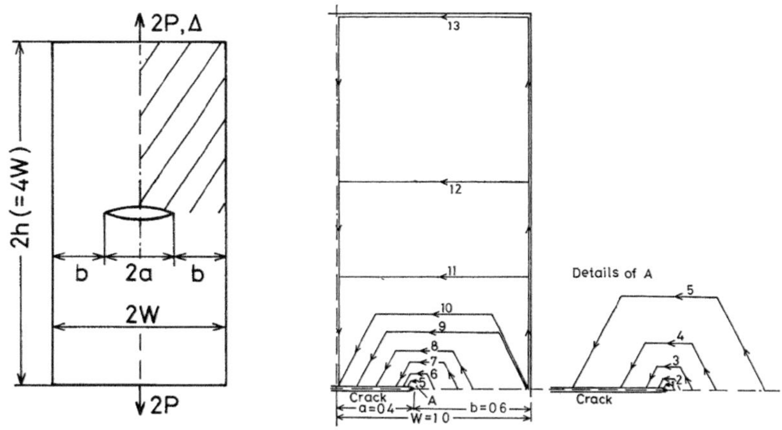

(a)

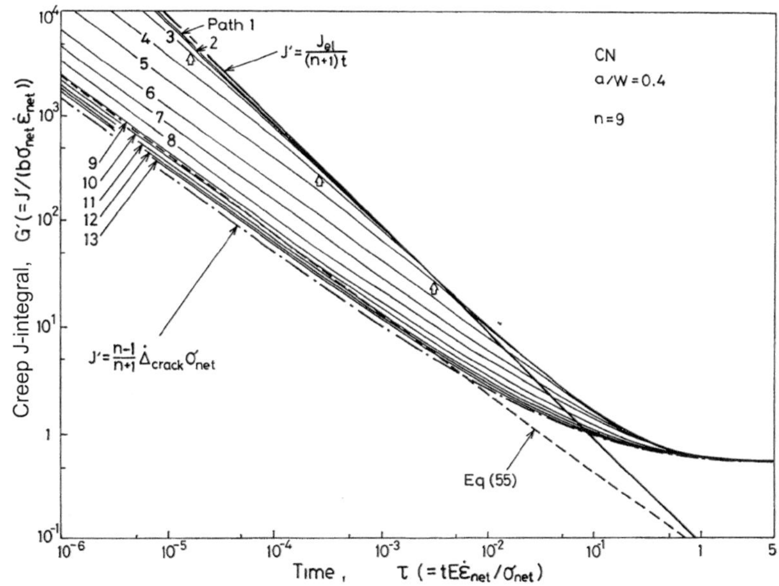

(b)

Figure 4.9 Change in the creep J-integral with loading time [Ohji *et al.*, 1980]. (a) Center-cracked plate and integral path of creep J-integral for finite element analysis. (b) Change in the creep J-integral.

The transition completion time is given by

$$t_{tr} = \frac{J_{el}}{(n+1)J^*_{st}} \tag{4.7}$$

as explained in Appendix 4-A. Furthermore, plastic deformation (non-linear deformation) at the loading instance promotes the transition, meaning that t_{tr} is shortened [Ohji *et al.*, 1980]. When the plastic deformation region is large, stress relaxation progresses near the crack tip, and the transition time is significantly shortened.

$C(t)$ parameter is proposed as the creep J-integral only for the integration paths within the creep-dominated area [Riedel and Rice, 1980]. As the creep-dominant region is confined to a very small area at $t \approx 0$, it can be applied to entire transition with the extreme integral path of $\Gamma \rightarrow 0$. In this book, including such meaning of $C(t)$, the integral is denoted as the creep J-integral, J^*.

4.5.2 Experiments in the Transient Condition

It is surprisingly difficult to experimentally verify the transition from small-scale creep to large-scale creep. There are only a limited number of reports in which crack propagation occurs during this transition. The difficulties stem from several factors:

(a) The completion time of the transition is too short in typical creep experiments.
(b) It takes a long time for crack initiation from the starter notch (or crack).
(c) Plastic deformation occurs around the crack tip at the loading instance.

Although creep brittle materials may present crack propagation under the transition, reliable experimental verification under static creep conditions is rare.

Creep crack propagation experiments are conducted at 1074 K on an intermetallic compound, TiAl, known for its low creep ductility [Tabuchi *et al.*, 2000]. However, the crack propagation rate, da/dt, is dominated by the creep J-integral, J^*, at the steady-state condition (post-transition). This indicates the difficulty of inducing the transition in laboratory settings.

In an investigation involving aluminum alloy 2014-T6, which exhibits even lower ductility, a notable relationship between da/dt and the linear elastic stress intensity factor, K, is observed, as depicted in Fig. 4.10 [Ohtani *et al.*, 1984]. This suggests that the state is quasi-static at $t \cong 0$, with crack propagates occurring while the singular stress field near the crack tip remains in the state of small-scale creep without the transition. Again, the characteristics during the transition have not been confirmed.

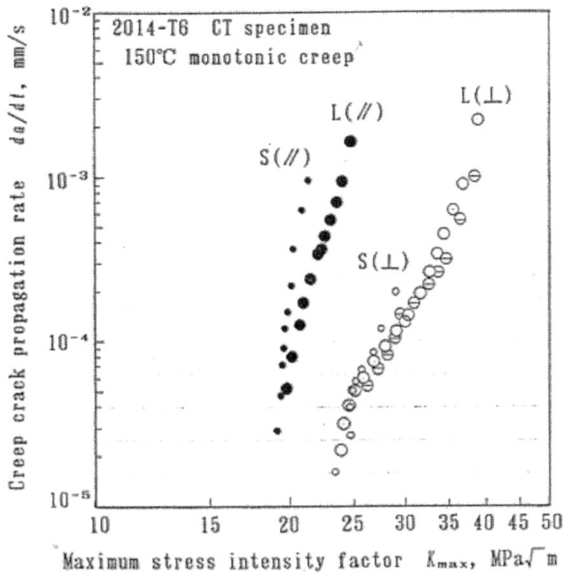

Figure 4.10 Crack propagation under the small-scale creep [Ohtani, 1984].

Carbon-fiber-reinforced polymer experiences delamination cracking between plies in creep at high temperatures, resulting in purely-time-dependent crack propagation. Due to the strong constraint imposed by the fibers on creep deformation of the polymer, the expansion of creep-dominated region is greatly limited. Consequently, small-scale creep persists for a long duration, with the linear elastic field governing the body. As depicted in Fig. 4.11, da/dt exhibits a good relationship with K (linear elastic energy release rate, G, in the figure) [Ohtani *et al.*, 1993]. This also fails to investigate the transition.

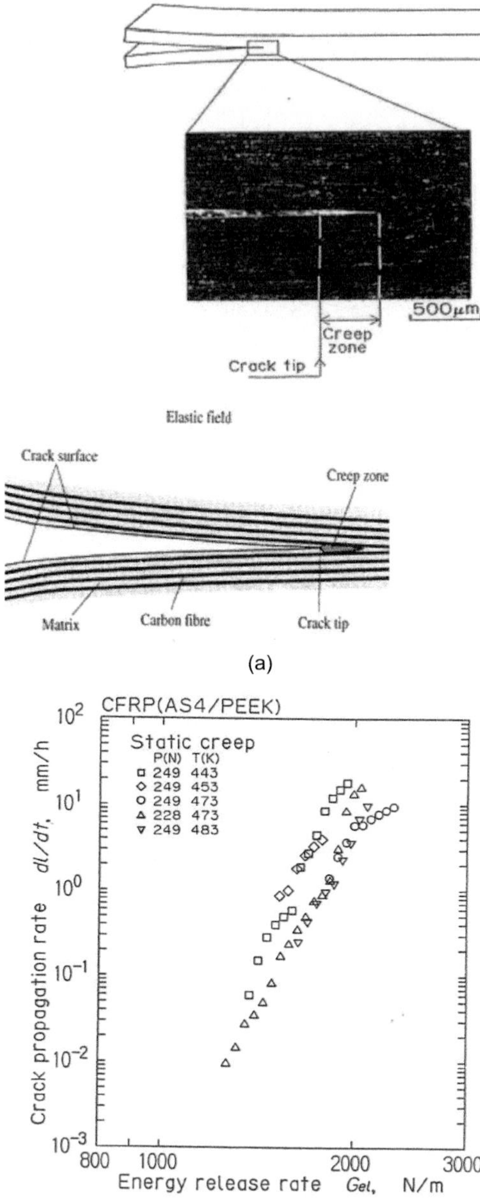

Figure 4.11 Creep crack propagation of carbon-fiber-reinforced polymer at high temperatures. Fiber: AS4, Matrix: Poly-Ether-Ether-Ketone [Ohtani *et al.*, 1993]. (a) Specimen of unidirectionally reinforced composite. (b) Relationship between the crack propagation rate and the (linear elastic) energy release rate.

Appendix

4-A Cracked Body under Interaction between Elasticity and Creep [Ohji *et al.*, 1980]

Consider an elastic and steady-state creep body with a crack subjected to a step load at $t = 0$. The total strain, ε, which is consisting of elastic strain, ε_{el}, and creep strain (steady-state creep) as

$$\varepsilon = \varepsilon_{el} + \int_0^t B\left[\sigma(s)\right]^n ds$$

$$= \frac{\sigma(t)}{E} + BF_s(t)\left[\sigma(t)\right]^n \quad (4A.1)$$

$$F_s(t) = \int_0^t \left[\frac{\sigma(s)}{\sigma(t)}\right]^n ds \quad (4A.2)$$

Assuming proportional loading, $F_s(t)$ is a function independent of the location in the body. Since Eq. (4A.1) is the summation of linear-elastic and power-law terms, it is analogous to the analysis used in the HRR singularity in Appendix 2-C (Chapter 2). Thus, the stress and strain fields are given by the following HRR-type equations

$$\sigma_{ij} = \left(\frac{J}{I_n BF_s(t) r}\right)^{\frac{1}{n+1}} \tilde{\sigma}_{ij}(\theta) \quad (4A.3)$$

$$\varepsilon_{ij} = BF_s(t)\left(\frac{J}{I_n BF_s(t) r}\right)^{\frac{n}{n+1}} \tilde{\varepsilon}_{ij}(\theta) \quad (4A.4)$$

As Eq. (4A.3) indicates that stress is proportional to $\left\{F_s(t)\right\}^{-\frac{1}{n+1}}$, the following relation is introduced by substituting into Eq. (4A.2),

$$F_s(t) = \int_0^t \left[\frac{F_s(t)}{F_s(s)}\right]^{\frac{n}{n+1}} ds \quad (4A.5)$$

Solving this, we have

$$F_s(t) = (n + 1)t \quad (4A.6)$$

Since linear elasticity dominates the body except the creep zone confined near the crack tip due to the small-scale creep condition, J

in Eq. (4A.3) is approximately equal to that in a linear elastic body J_{el}, considering the path independence. Then, the stress field near the crack tip is in the following form, substituting them into Eqs. (4A.3) and (4A.4):

$$\sigma_{ij} = \left(\frac{J_{el}}{I_n B(n+1)tr} \right)^{\frac{1}{n+1}} \tilde{\sigma}_{ij}(\theta) \tag{4A.7}$$

$$\dot{\varepsilon}_{ij} = B \left(\frac{J_{el}}{I_n B(n+1)tr} \right)^{\frac{n}{n+1}} \tilde{\varepsilon}_{ij}(\theta) \tag{4A.8}$$

On the other hand, since the zone near the crack tip is governed by steady-state creep, the HRR singular fields, Eqs. (3.12) and (3.13), hold. Equating those, the creep J-integral is related to the J-integral (and the linear elastic stress intensity factor, K):

$$J^* = \frac{J_{el}}{(n+1)t} \qquad (4A.9) = (4.6)$$

It should be noted that J^* is the value evaluated on the integration path inside the creep-dominated zone, where the creep strain is much larger than the elastic one. Although J^* is path-independent in the zone, it is not outside it.

The transition time, t_{tr}, from small-scale creep to large-scale creep is approximately given by equating J^*_{st} in the steady state where the creep strain dominates the entire body, and J^* in the transient condition, Eq. (4A.9) as

$$t_{tr} = \frac{J_{el}}{(n+1)J^*_{st}} \qquad (4A.10) = (4.7)$$

References

Kubo, S., 1983, Effects of creep recovery and hardening on the stress and strain-rate fields near a crack tip in creeping materials, ASTM STP 803, pp. I594–I614.

Ohji K., K. Ogura, and S. Kubo, 1980, Stress field and modified J-integral near a crack tip under condition of confined creep deformation, *Journal of the Society of Materials Science Japan*, 29, pp. 465–471. https://www.jstage.jst.go.jp/article/jsms1963/29/320/29_320_465/_pdf/-char/ja

Ohtani, R., K. Yamada, T. Kashiwagi, and H. Matsubara, 1982, Crack propagation of high temperature fatigue in 304 stainless steel, *Transaction of Japanese Society of Mechanical Engineers*, 48, pp. 1378–1390. https://www.jstage.jst.go.jp/article/kikaia1979/48/435/48_435_1378/_pdf/-char/ja

Ohtani, R., J. Nishimoto, and R. Hiratsuka, 1984, Measurement of crack length at elevated temperature using D.C. electrical potential method, *Journal of the Society of Materials Science Japan*, 33, pp. 729–735. https://www.jstage.jst.go.jp/article/jsms1963/33/369/33_369_729/_pdf/-char/ja

Ohtani, R., T. Kitamura, and S. Enomoto, 1988, Creep crack propagation under decreasing load condition, *Journal of the Society of Materials Science Japan*, 37, pp. 283–288. https://www.jstage.jst.go.jp/article/jsms1963/37/414/37_414_283/_pdf/-char/ja

Ohtani R., T. Kitamura, Y. Uematsu, and M. Hojo, 1993, Mode I delamination of carbon fiber reinforced thermoplastic polymer under static and cyclic creep at elevated temperatures, ASTM STP 1174, pp. 23–38.

Riedel, H., and J.R. Rice, 1980, Tensile cracks in creeping solids, ASTM STP 700, pp. 112–130.

Riedel, H., 1981, Creep deformation at crack tips in elastic-viscoplastic solids, *Journal of Mechanical Physics of Solids*, 29, pp. 35–49.

Saxena, A., 1986, Creep crack growth under non-steady-state conditions, ASTM STP 905, pp. 185–201.

Tabuchi, M., T. Yokobori, A. Fuji, K. Kubo, K. Yagi, and T. Yokobori, 2000, Creep crack growth test method and creep crack growth properties for TiAl intermetallic compound, *Journal of the Society of Materials Science Japan*, 49, pp. 80–85. https://www.jstage.jst.go.jp/article/jsms1963/49/1/49_1_80/_pdf/-char/ja

Taira, S., R. Ohtani, and T. Komatsu, 1979, Application of J-integral to high-temperature crack propagation, Part II – fatigue crack propagation, *Journal of Engineering Materials and Technology*, 101, pp. 162–167.

Chapter 5

Cycle Dependence and Time Dependence in Crack Propagation

5.1 Two Different Mechanisms of Crack Propagation at High Temperatures

At low temperatures where creep does not occur, repeated plastic strain (crystallographic slips) near the crack tip brings about crack propagation, namely fatigue crack propagation. The typical mechanism proposed is blunting and re-sharpening of crack tip [Laired, 1967]. This indicates that crack propagation is purely cycle-dependent and is governed by the range of strain (stress) intensity near the crack tip, as explained in Section 2.3.2 (Chapter 2). Since the mechanism works under cyclic loading at high temperatures as well, there is cycle-dependent crack propagation in high temperature fatigue.

On the other hand, as shown in Section 3.3 (Chapter 3), creep brings about crack propagation that is purely time-dependent and is governed by the strain rate (stress) near the crack tip. The mechanism might involve creep damage, such as small voids and micro cracks concentrated near the crack tip. This has completely different characteristics from the above fatigue (cycle-dependent) crack propagation. Moreover, this crack propagation mechanism operates even under cyclic stress changes if the body is stressed for a long time, as discussed in Chapter 4.

High Temperature Fracture Mechanics
Takayuki Kitamura
Copyright © 2025 Jenny Stanford Publishing Pte. Ltd.
ISBN 978-981-5129-55-7 (Hardcover), 978-1-003-65328-8 (eBook)
www.jennystanford.com

In the early stages of research on crack propagation at high temperatures, focusing on the significant variations in the crack propagation rates depending on the factors of time, cycles, stress, temperature was investigated based on diverse ideas. [ASTM STP 520, 1973]. Among those trials, the following expression clearly captures the essential feature.

Figure 5.1 schematically depicts diagrams showing the dependence of crack propagation rate on stress-cycle frequency, ν, under the same stress conditions (both mean and amplitude) at the same crack length (for example, [Solomon and Coffin,1973]). It should be noted that the crack propagation rate has two forms—crack extension per unit cycle, $\mathrm{d}a/\mathrm{d}N$, and per unit time, $\mathrm{d}a/\mathrm{d}t$, where $\mathrm{d}a/\mathrm{d}t$ is converted by dividing $\mathrm{d}a/\mathrm{d}N$ by the tensile period, $1/(2\nu)$. Figures 5.1(a) to (d) represent different expressions of the same dataset to facilitate comprehension of the key feature. The relationship embodies fundamental characteristics that offer crucial insights into understanding the entirety of fracture mechanics in high temperature fatigue [Taira *et al.*, 1979].

The relationship exhibits three regions:

(1) High frequency region

Due to insufficient loading time to generate creep strain, only elastoplastic strain occurs in the cycles. Consequently, the crack propagation rate per unit cycle, $\mathrm{d}a/\mathrm{d}N$, remains constant regardless of ν without the influence of time—purely cycle-dependent fatigue.

(2) Low frequency region

As loading time is long enough to yield creep strain, it prevails elastoplastic strain in the cracked body. This results in the crack propagation rate per unit cycle, $\mathrm{d}a/\mathrm{d}N$, being inversely proportional to ν, indicating that the crack propagation rate per unit time, $\mathrm{d}a/\mathrm{d}t$, remains constant in the region—purely time-dependent fatigue. This region is characterized by (steady-state) creep-dominated crack propagation.

(3) Medium frequency region

It is considered that both mechanisms simultaneously operate in this region. It should be noted that the crack propagation rate is significantly faster than the summation of rates in regions (1) and (2). This suggests an interaction between the two mechanisms, namely the crack propagation region under creep–fatigue interaction. Fracture mechanics in this region is explained in Chapter 7.

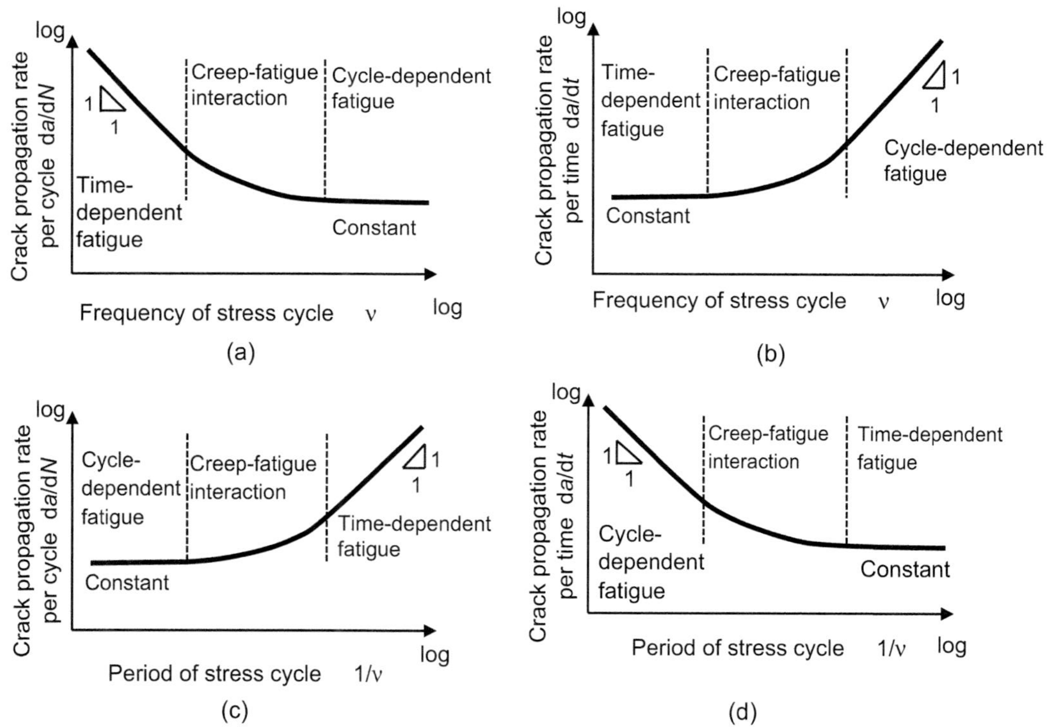

Figure 5.1 Schematic diagrams showing influence of cycle-dependence and time-dependence on crack propagation in high temperature fatigue. (a) Relationship between the crack propagation rate per cycle, da/dN, and the stress-cycle frequency, ν. (b) Relationship between the crack propagation rate per unit time, da/dt, and the stress-cycle frequency, ν. (c) Relationship between the crack propagation rate per cycle, da/dN, and the stress-cycle period, $1/\nu$. (d) Relationship between the crack propagation rate per unit time, da/dt, and the stress-cycle period, $1/\nu$.

5.2 Cycle-Dependent Crack Propagation (Pure Fatigue)

5.2.1 Fracture Mechanics Law of Fatigue Crack Propagation

As explained in Section 2.3.2 (Chapter 2), the effective linear elastic stress intensity factor range, ΔK_{eff}, governs fatigue crack propagation under elasticity-dominated conditions such as small-scale yielding (high-cycle fatigue). Extending to elastoplastic conditions (low-cycle fatigue), the fatigue J-integral range, ΔJ_f, emerges a reasonable parameter for cycle-dependent crack propagation.

Figure 5.2 shows experimental results of crack propagation of carbon steel in cycle-dependent fatigue with a frequency of 1 Hz at 673 K [Ohtani and Kitamura, 1988]. In this condition, creep strain is either absent or negligible due to the short tensile period. Fracture surface morphology resembles that observed in room temperature fatigue.

The da/dN–ΔK relationship shown in Fig. 5.2(a) exhibits a power law in the middle ΔK range and the existence of threshold at low ΔK. As explained in Chapter 2, near the threshold, ΔK_{eff} provides better correlation, which is also plotted in Fig. 5.2(a). Thus, the crack propagation law in high-cycle fatigue is summarized as

$$\frac{da}{dN} = c_f \Delta K_{\text{eff}}^{\,m_{kf}} \qquad (5.1) = (2.19)$$

except for the threshold.

In case of high loading, a notable acceleration is observed compared to the relationship described in Eq. (5.1), resulting in the loss of effectiveness of the linear elastic stress intensity factor as shown by solid marks in Fig. 5.2(a). Figure 5.2(b) represents the same data, but plotted as da/dN against ΔJ_f, which is evaluated using a graphical method based on the area in the load-displacement curve (Eq. (2.25) and Fig. 2.9). Regardless of whether the fatigue is elastic or elastoplastic, there exists a good correlation between da/dN and ΔJ_f as

$$\frac{da}{dN} = \Delta J_f^{\,m_f} \quad \text{when} \quad \Delta J_f \geq \Delta J_{f\,th} \qquad (5.2) = (2.24)$$

Simplifying the transient behavior near the threshold

$$\frac{da}{dN} = 0 \quad \text{when} \quad \Delta J_f < \Delta J_{f\,th} \qquad (5.3) = (2.26)$$

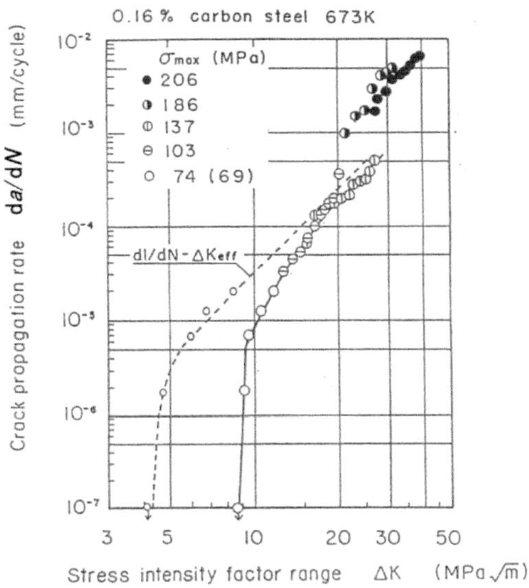

(a)

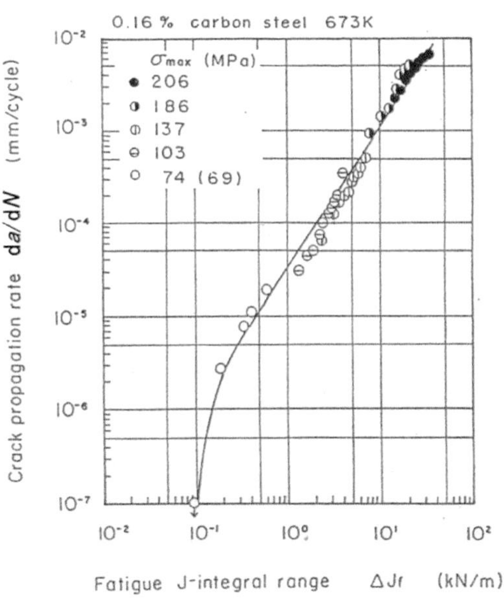

(b)

Figure 5.2 Crack propagation in cycle-dependent fatigue of carbon steel at 673 K [Ohtani and Kitamura, 1988]. (a) Relationship between the crack propagation rate and the linear elastic stress intensity factor range. The dashed line presents the crack propagation rate as a function of the effective linear elastic stress intensity factor, ΔK_{eff}. (b) Relationship between the crack propagation rate and the fatigue J integral range, ΔJ_f.

5.2.2 Influence Factors

(1) Cycle period

Figure 5.3 shows the dependence of da/dN on cycle period, $1/\nu$, in fatigue of 304 stainless steel at 823 K under the same stress condition and at the same crack length [Kitamura, 1986]. da/dN remains almost constant in the region with shorter period ($1/\nu < 1000$ seconds), indicating purely cycle-dependent crack propagation.

(2) Temperature

Figure 5.4 shows experimental results in fatigue of 304 stainless steel under the same loading condition except for only temperature [Kitamura, 1986]. Although da/dN increases at higher temperatures, this is attributed to the increase in plastic strain resulting from the decreased yield strength.

Figure 5.5 depicts the da/dN–ΔJ_f relationship obtained from the experiments presented in Figs. 5.3 and 5.4. Regardless of cycle period and temperature, crack propagation is uniquely governed by ΔJ_f [Kitamura, 1986].

(3) Compressive creep

It has been reported that the failure life of smooth specimens in high temperature fatigue is strongly influenced by compressive creep strain, known as the strain (stress) waveform effect [Manson, 1973; Halford *et al.*, 1973]. This suggests that the crack propagation rate might be affected by compressive creep, a kind of creep–fatigue interaction. On the other hand, since ΔJ_f is evaluated solely from the tension-going portion in a cycle, this suggests no influence of compressive creep strain on crack propagation.

To investigate this contradicting influence, Fig. 5.6 [Kitamura, 1986] presents the relationships between da/dN and ΔJ_f in fatigue of Ni-Mo-V steel at 823 K without compression creep (indicated as "pp" in the figure) and with it (indicated as "pc"), respectively. No prominent difference is observed, indicating crack propagation primarily takes place in the tension-going half cycle [Ohtani *et al.*, 1982].

It should be noted that compressive creep may alter the plastic behavior (constitutive equation) in tension. In this case, it is automatically included through the evaluation based on P–δ curve in the magnitude of ΔJ_f.

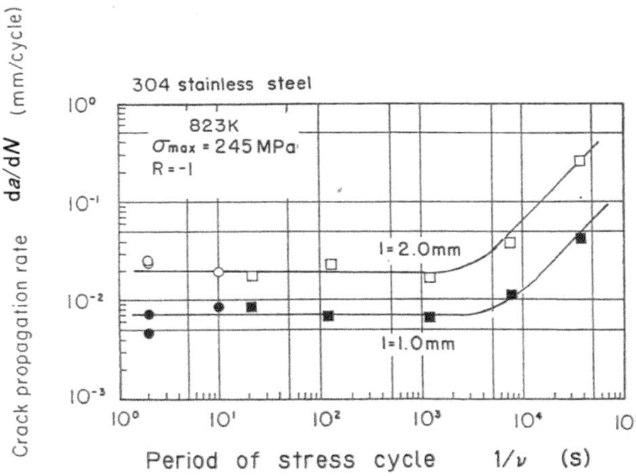

Figure 5.3 Dependence of the crack propagation rate, d*a*/d*N*, on the cycle period, $1/\nu$, in cycle-dependent fatigue of 304 stainless steel at 823 K under the crack length of 1 mm (solid mark) and 2 mm (hollow mark) [Kitamura, 1986].

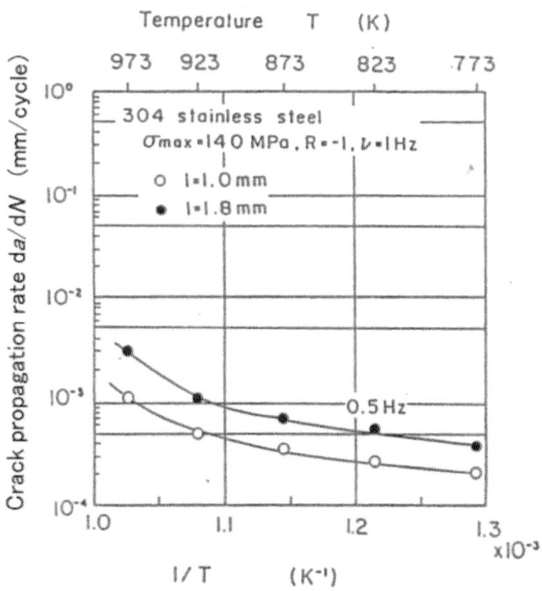

Figure 5.4 Dependence of the crack propagation rate, d*a*/d*N*, on the temperature, *T*, in cycle-dependent fatigue of 304 stainless steel at cycle frequency of 1 Hz (0.5 Hz at 823 K) under the crack length of 1 mm (solid mark) and 2 mm (hollow mark) [Kitamura, 1986].

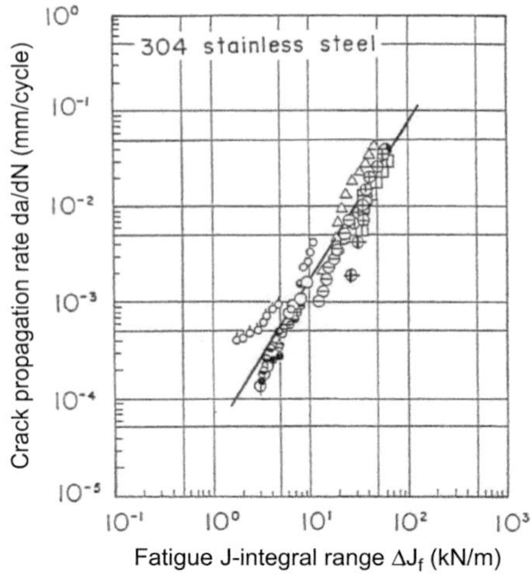

Figure 5.5 Relationship between the crack propagation rate, d*a*/d*N*, and the fatigue J-integral range, ΔJ_f, in the experiments of Figs. 5.3 and 5.4 [Kitamura, 1986].

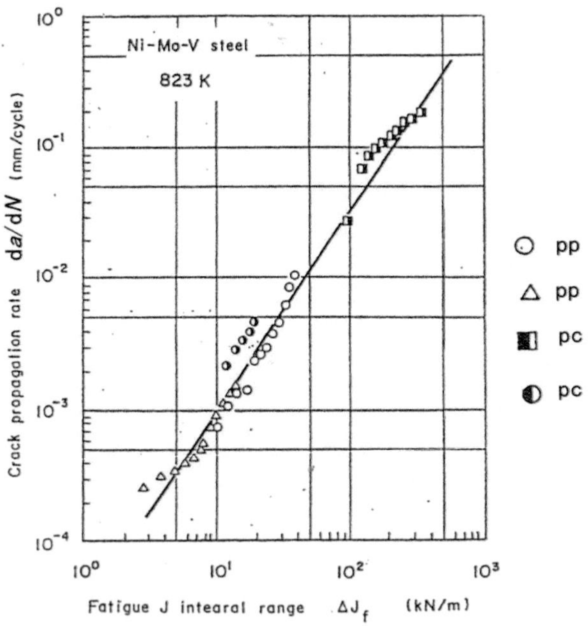

Figure 5.6 Influence of compressive creep on the relationship between the crack propagation rate, d*a*/d*N*, and the fatigue J integral, ΔJ_f in cycle-dependent fatigue. "pp" is the fatigue without compressive creep while "pc" is the one with it [Kitamura, 1986].

5.2.3 Cycle-Dependent Crack Propagation of Various Steels and Alloys

Figure 5.7 presents a summary [Ohtani and Kitamura, 1988] of crack propagation obtained in the author's laboratory in cycle-dependent fatigue of various materials, demonstrating that the relationships fall within a narrow band.

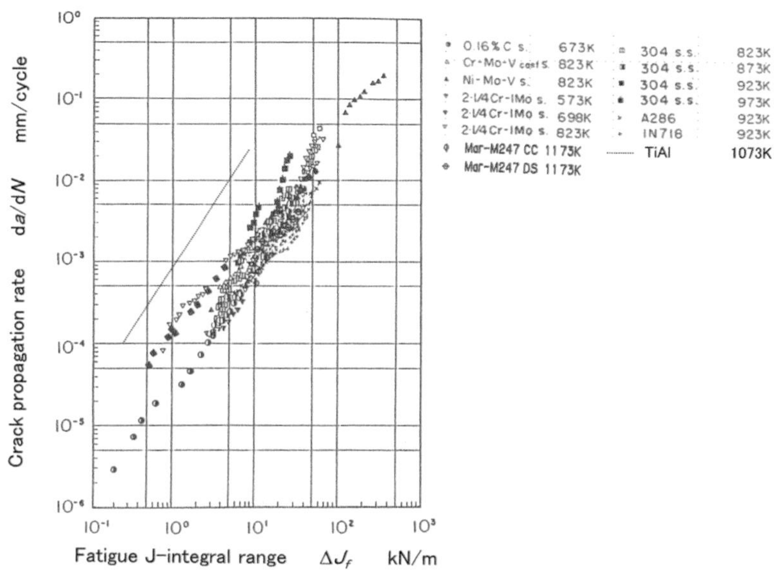

Figure 5.7 Crack propagation properties in cycle-dependent fatigue of various alloys and steels [Kitamura, 1986].

5.3 Time-Dependent Crack Propagation (Pure Creep)

5.3.1 Crack Propagation in Cyclic Creep

The fundamental behavior of time-dependent crack propagation in high temperature fatigue can be investigated through experiments using a trapezoidal waveform with holds at maximum and minimum stress with fully reversed loading. This condition, known as cyclic creep, is convenient for extracting the creep effect in fatigue loading [Taira *et al.*, 1979 and 1980]. Since deformation during the tensile stress hold is caused by creep, the crack center displacement rate, $\dot{V}$, is measured for evaluation of the creep J-integral using Eq. (3.15). While quantifying the crack propagation rate during the stress hold is challenging, the crack propagation rate per unit cycle, d*a*/d*N*, is easily evaluated from the *a*–*N* curve. The crack propagation rate per unit time, d*a*/d*t*, is then converted from d*a*/d*N* by dividing by the stress hold period as an average value.

Figure 5.8 shows the relationship between d*a*/d*t* and *J** in cyclic creep of carbon steel at 673 K [Taira *et al.*, 1980], including the ones in static creep. The good agreement indicates the same mechanism governs crack propagation static creep and time-dependent fatigue (cyclic creep).

$$da/dt = C_c J^{*m} \cong C_c J^* \qquad (5.4) = (3.16)$$

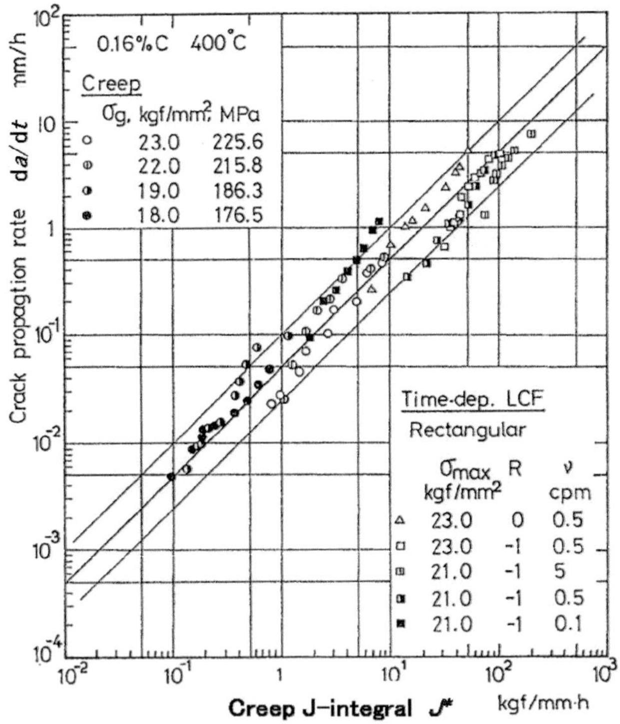

Figure 5.8 Applicability of creep J-integral on crack propagation in cyclic creep showing the relationship between the crack propagation rate, d*a*/d*t*, and the creep J-integral, *J**, in cyclic creep of carbon steel at 673 K [Taira *et al.*, 1980].

5.3.2 Crack Propagation in Strain-Controlled Fatigue

Time-dependent crack propagation takes place not only in cyclic (stress-controlled) creep but also in strain-controlled fatigue with a slow tensile strain rate. Considering that creep strain is approximately equal to inelastic strain in fatigue with a slow strain rate and long period, the magnitude of J^* can be evaluated by

$$\dot{\varepsilon}_{in} = \frac{\Delta\varepsilon_{in}}{\left(1/2v\right)} \tag{5.5}$$

where $\Delta\varepsilon_{in}$ is the inelastic strain range.

Figure 5.9 shows the relationship between da/dt and J^* for time-dependent crack propagation in strain-controlled fatigue of Hastelloy X at 1073 K [Taira *et al.*, 1980], where temperature and strain rate conditions sufficiently meet the above requirement [Ohji *et al.*, 1981]. Since the relationship between $\dot{\varepsilon}_{in}$ and σ aligns well with the Norton law in static creep (Eq. (3.1)), steady-state creep governs the deformation. The figure clearly demonstrates a strong correlation between da/dt and J^*.

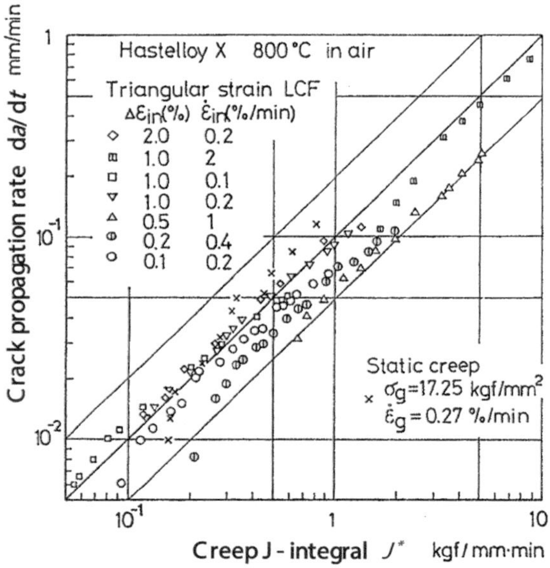

Figure 5.9 Relationship between the crack propagation rate, da/dt, and the creep J-integral, J^*, in strain-controlled time-dependent fatigue (low frequency) of Hastelloy X at 1073 K [Taira *et al.*, 1980].

References

ASTM STP 520, 1973, *Fatigue at Elevated Temperatures.*

Halford, G.R., M.H. Hirschberg, and S.S. Manson, 1973, Temperature effects on the strainrange partitioning approach for creep fatigue analysis, ASTM STP, pp. 658–669.

Jeglic F., P. Niessen, and D.J. Burns, Temperature dependence of fatigue crack propagation in an Al–2.6Mg Alloy, pp. 139–148.

Kitamura, T., 1986, Fracture Mechanics Investigation on Crack Propagation in High Temperature Fatigue, Doctoral Thesis.

Laird, C., 1967, The influences of metallurgical structure on the mechanisms of fatigue crack propagation, ASTM STP 415, pp. 136–180.

Manson S.S., 1973, The challenge to unify treatment high temperature fatigue—A partisan proposal based on strainrange partitioning, ASTM STP 520, pp. 744–782.

McHenry, M.I., and A.W. Pense, Fatigue crack propagation in steel alloys at elevated temperatures, pp. 345–354.

Ohtani, R., and T. Kitamura, 1988, Characterization of high temperature strength of metals based on the mechanics of crack propagation, In *High Temperature Creep-Fatigue*, R. Ohtani, M. Ohnami and T. Inoue (eds.), Elsevier Applied Science, London, 65–90.

Ohtani, R., K. Yamada, T. Kashiwagi, and H. Matsubara, 1982, Crack propagation of high-temperature fatigue in 304 stainless steel, *Transaction of Japanese Society of Mechanical Engineers*, 48, pp. 1378–1390. https://www.jstage.jst.go.jp/article/kikaia1979/48/435/48_435_1378/_pdf/-char/ja

Pook, L.P., and A.A. Beveridge, Threshold for fatigue crack growth in ferritic steel at 300 C, pp. 179–191.

Rau, C.A., A.E. Gemma, and G.R. Leverant, Thermal-mechanical fatigue crack propagation in nickel- and cobalt-base superalloys under various strain-temperature cycles, pp. 166–178.

Shahinian, R., H.H. Smith, and H.E. Watson, Fatigue crack growth characteristics of several austenitic stainless steels at high temperature, pp. 387–400.

Solomon, H.D., and Coffin, L.F. Jr, 1973, Effects of frequency and environment on fatigue crack growth in A286 at 1100 F, ASTM STP 520, pp. 112–122.

Taira, S., R. Ohtani, and T. Komatsu 1979, Application of J-integral to high-temperature crack propagation, Part II Fatigue crack propagation, *Journal of Engineering Materials and Technology*, 101, pp. 162–167.

Taira, S., R. Ohtani, T. Yonekura, M. Osada, and T. Kitamura 1980, Time-dependent fatigue crack propagation at high temperatures (in Japanese), *Transaction of Japanese Society of Mechanical Engineers A*, 46, pp. 861–869. https://www.jstage.jst.go.jp/article/kikaia1979/46/408/46_408_861/_pdf/-char/ja Ohji, K.K. Ogura, and S. Kubo, (in Japanese), Study on simplified evaluation equations of j-integral and modified J-integral, *Transaction of Japanese Society of Mechanical Engineers A*, 47, pp. 400–405. https://www.jstage.jst.go.jp/article/kikaia1979/47/416/47_416_400/_pdf/-char/ja

Chapter 6

Transient Crack Propagation in Time-Dependent Fatigue

6.1 Creep J-Integral Range

6.1.1 Creep J-Integral in a Cycle

As explained in Chapter 4, transient acceleration (or deceleration) of crack propagation occurs after a stress change. Since fatigue loading consists of cyclic stress changes, this transition takes place in every cycle, resulting in more significant influence on crack propagation in creep-dominated (time-dependent) fatigue. Fortunately, the results presented in Chapter 4 suggest that the effect of a simple stress change on creep crack propagation can be assessed by the creep J-integral, which includes the transient acceleration of creep rate near the crack tip. This implies the applicability of the creep J-integral to crack propagation in time-dependent fatigue.

The relationship between $\mathrm{d}a/\mathrm{d}t$ and $\mathrm{d}a/\mathrm{d}N$ is expressed as follows:

$$\frac{da}{dN} = \int_0^{t_1} \frac{da}{dt} dt \tag{6.1}$$

High Temperature Fracture Mechanics
Takayuki Kitamura
Copyright © 2025 Jenny Stanford Publishing Pte. Ltd.
ISBN 978-981-5129-55-7 (Hardcover), 978-1-003-65328-8 (eBook)
www.jennystanford.com

where t_1 is the period of tensile half cycle. The average crack propagation rate, $\left(\frac{da}{dt}\right)_{av}$, is defined as

$$\left(\frac{da}{dt}\right)_{av} = \frac{da/dN}{t_1} \tag{6.2}$$

Referring Eq. (6.1), the creep J-integral range, ΔJ_c, is defined as the accumulated amount of the creep J-integral in a unit cycle [Taira *et al.*, 1979], which is formulated as

$$\Delta J_c = \int_0^{t_1} J^*(t)dt \tag{6.3}$$

As a similar expression of Eq (6.2), the average creep J-integral, $(J^*)_{av}$ is evaluated by [Kuwabara *et al.*, 1988]

$$\left(J^*\right)_{av} = \frac{\Delta J_c}{t_1} \tag{6.4}$$

Since the fracture mechanics law in creep crack propagation is

$$\frac{da}{dt} \cong C_c J^* \qquad (6.5) = (3.16)$$

we have the following laws:

$$\frac{da}{dN} = C_c \Delta J_c \tag{6.6}$$

$$\left(\frac{da}{dt}\right)_{av} = C_c \left(J^*\right)_{av} \tag{6.7}$$

6.1.2 Experimental Evaluation of Creep J-Integral

The constitutive equation (stress and strain/strain rate relationship) is highly sensitive to stress changes in time-dependent (creep-dominated) fatigue, resulting in an extremely complex form. Fortunately, the dependence of the experimental evaluation method, Eq. (3.15) [Ohji *et al.*, 1978], on the constitutive equation is weak

$$J^* \cong \sigma_{net} \dot{V} \qquad (6.8) = (3.15)$$

Then, by measuring the creep component of crack opening displacement, V, which is easily observed at each instance in a cycle of cyclic creep, the change in the transient creep J-integral, $J^*(t)$, can be evaluated. The creep J-integral range is evaluated as

$$\Delta J_c \cong \sigma_{net} \Delta V \tag{6.9}$$

using the crack center opening displacement accumulated during stress hold, ΔV, in cyclic creep.

6.2 Acceleration due to Transient Creep

6.2.1 Acceleration of Creep J-Integral [Kitamura *et al.*, 1986]

The Norton law in steady-state creep of 0.16% carbon steel at 673 K is formulated as

$$\dot{\varepsilon} = 9.43 \times 10^{-35} \sigma^{15.4} \tag{6.10}$$

where the units are ($\dot{\varepsilon}$) h^{-1} – (σ) MPa. The magnitude of J^* of a center-cracked plate specimen in steady-state creep can be analytically evaluated based on Eq. (6.10) [Ohji *et al.*, 1981]. This creep J-integral is denoted as J^*_{st} and the creep J-integral range is simply calculated by $(\Delta J_c)_{st} = J^*_{st} t_H$, where t_H is the stress hold period. Crack propagation experiments are conducted in cyclic creep of this 0.16% carbon steel at 673 K with a stress ratio of $R = \sigma_{min}/\sigma_{max} = -1$ and t_H = 60 seconds, and the relationship between da/dN and $(\Delta J_c)_{st}$ is shown in Fig. 6.1(a) [Kitamura *et al.*, 1986]. It is evident that da/dN does not exhibit a good correlation with $(\Delta J_c)_{st}$. Furthermore, compared to results in static creep (converted from da/dt–J^*_{st} to da/dt–$(\Delta J_c)_{st}$), crack propagation is significantly accelerated. This indicates that transient creep expedites crack propagation by several orders of magnitude.

On the other hand, the creep J-integral range is experimentally evaluated by Eq. (6.9) measuring the change in crack opening displacement, V, denoted as $(\Delta J_c)_{ex}$ to distinguish it from $(\Delta J_c)_{st}$. Figure 6.1(b) shows the comparison between $(\Delta J_c)_{ex}$ and $(\Delta J_c)_{st}$. As the former incorporates the effect of transient creep, this clearly demonstrates that the creep J-integral is also significantly accelerated. Since the carbon steel is creep-ductile, this acceleration is attributed to the material factor, transient creep.

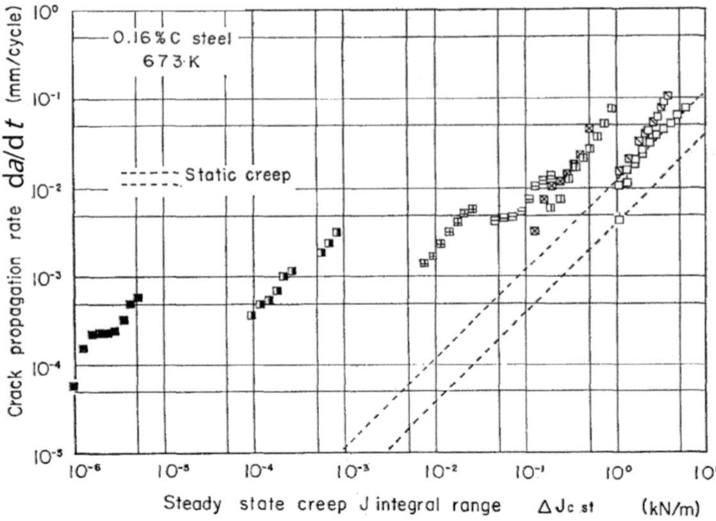

(a)

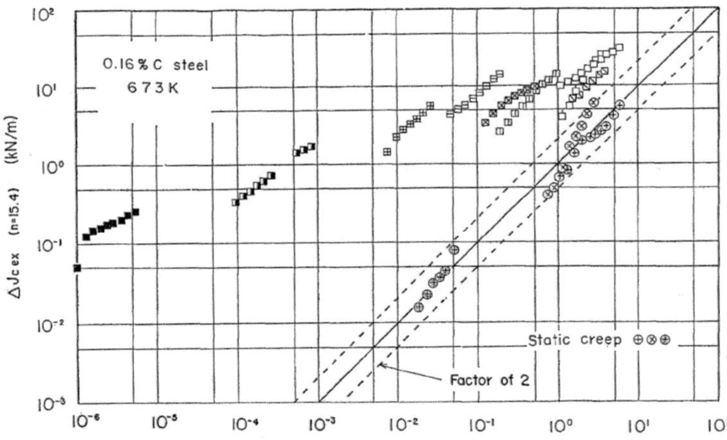

(b)

Figure 6.1 The steady-state creep J-integral range in cyclic creep of 0.16% carbon steel at 673 K with $t_H = 60$ seconds, $(\Delta J_c)_{st}$, which is analytically evaluated using the constitutive equation in steady-state creep (Eq. (6.10)) [Kitamura *et al.*, 1986]. (a) Relationship between the crack propagation rate, d*a*/d*N*, and the steady-state creep J-integral, $(\Delta J_c)_{st}$. The dashed-line band show the relationship in static creep. (b) Comparison between the steady-state creep J-integral range, $(\Delta J_c)_{st}$, and the experimentally evaluated creep J-integral range, $(\Delta J_c)_{ex}$. The dashed-line band show the relationship in static creep ($t_H = 60$ seconds for static creep).

Figure 6.2(a) shows the $\dot{\varepsilon}$ (h^{-1}) – σ (MPa) relationship obtained from smooth specimens of the carbon steel without crack/notch during tensile stress hold in cyclic creep with $R = \sigma_{min}/\sigma_{max} = -1$ at 673 K. At any stress level, the strain rate is initially fast after loading, and subsequently decreases with time, which is a typical feature of

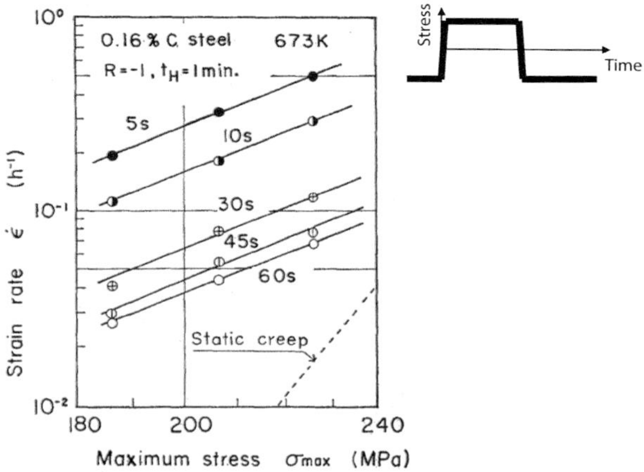

(a)

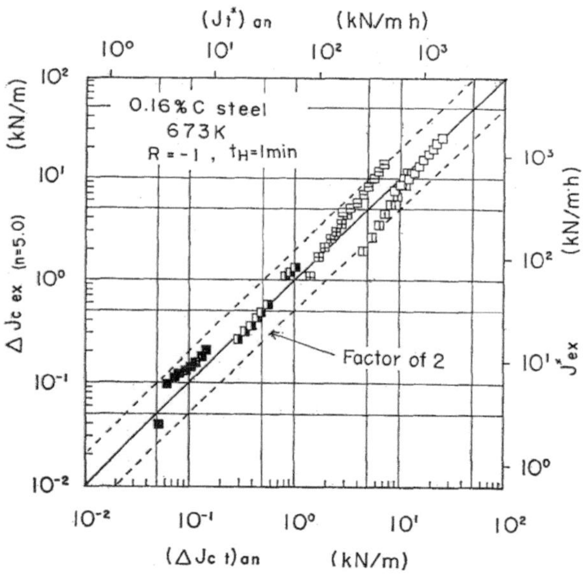

(b)

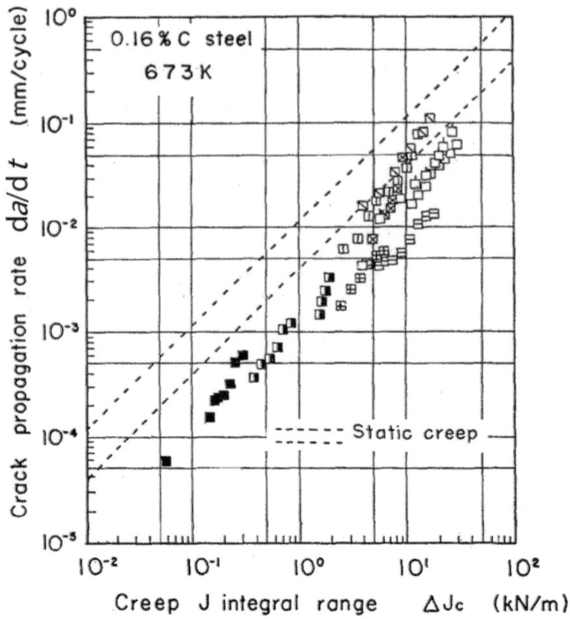

(c)

Figure 6.2 The creep J-integral range in cyclic creep of 0.16% carbon steel at 673 K taking into account the acceleration due to the constitutive relation in transient creep [Kitamura *et al.*, 1986]. (a) Dependence of strain rate on hold tensile stress in cyclic creep of 0.16% carbon steel at 673 K and change of strain rate during hold period. The applied stress is schematically shown in the figure. (b) Comparison between the analytically evaluated transient state creep J-integral range based on Eq. (6.11), $(\Delta J_c)_{an}$, and the experimentally evaluated creep J-integral range, $(\Delta J_c)_{ex}$. (c) Relationship between the crack propagation rate, d*a*/d*N*, and the experimentally evaluated creep J-integral range, $(\Delta J_c)_{ex}$. The dashed-line band show the relationship in static creep.

transient creep. The constitutive equation in steady-state creep (Eq. (6.10)) is depicted by the dashed line in the figure, presenting that the strain rate does not reach steady-state even at the end of tensile stress hold period. The accumulated creep strain during the tensile stress hold period (60 seconds) is formulated with respect to stress as

$$\Delta\varepsilon_c = \int_0^{t_H} \dot{\varepsilon} dt = 7.564 \times 10^{-15} \sigma_{\max}^{\ 5.0} \tag{6.11}$$

Similarly to $(\Delta J_c)_{st}$, the creep J-integral range in the transition can be analytically evaluated based on Eq. (6.11), denoted as $(\Delta J_c)_{an}$. As shown in Fig. 6.2(b), the experimentally evaluated $(\Delta J_c)_{ex}$ correlates well with the analytically evaluated value based on transient creep $(\Delta J_c)_{an}$, suggesting that the creep J-integral is well assessed by the experimental method encompassing transient condition.

Both of da/dN and ΔJ_c are equally accelerated during the transition, resulting in a good correlation between da/dN and $(\Delta J_c)_{ex}$, as shown in Fig. 6.2(c). Here, Fig. 6.2(c) illustrates that cyclic creep exhibits lower propagation rate than static creep (as indicated by the dashed line band in the figure) for the same magnitude of creep J-integral range. This discrepancy is attributed to the characteristic complex phenomena arising from creep–fatigue interaction caused by the presence of compressive creep, which is explained in Section 8.1.

6.2.2 Acceleration of Strain Rate Near the Crack Tip [Taira *et al.*, 1980]

Figure 6.3(a) shows the crack propagation rate in the cyclic creep of 0.16% carbon steel at 673 K under constant maximum stress with various stress ratio. Transient creep significantly accelerates da/dt, and a lower minimum stress brings about a higher propagation rate. The strain near the crack tip, ε_{ct}, is observed during the tensile stress hold by a fine mesh, and Fig. 6.3(b) demonstrates the change in ε_{ct}, revealing that strain rate tends to stabilize in the latter stage of the hold period. The strain rate at this stage is selected as representative, and its magnitudes are plotted in Fig. 6.3(c) The strain rate near the crack tip is greatly accelerated by transient creep, exhibiting the same tendency as that of da/dt, shown in Fig. 6.3(a).

Figure 6.3(d) shows the relationship between da/dt and $\dot{\varepsilon}_{ct}$. Regardless of the stress ratio (degree of recovery), the results exhibit the same relationship and coincide with those in static creep, meaning that fracture driving force is governed by J^*. This also indicates the same mechanism of crack propagation.

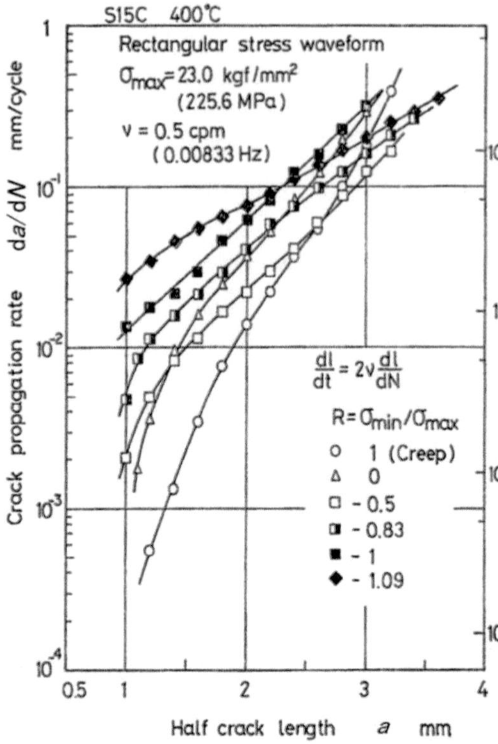
S15C 400°C
Rectangular stress waveform
σmax=23.0 kgf/mm²
(225.6 MPa)
ν = 0.5 cpm
(0.00833 Hz)
dl/dt = 2ν dl/dN
R=σmin/σmax
1 (Creep)
0
- 0.5
- 0.83
- 1
- 1.09
Crack propagation rate da/dN mm/cycle
dl/dt, mm/h
Half crack length a mm

(a)

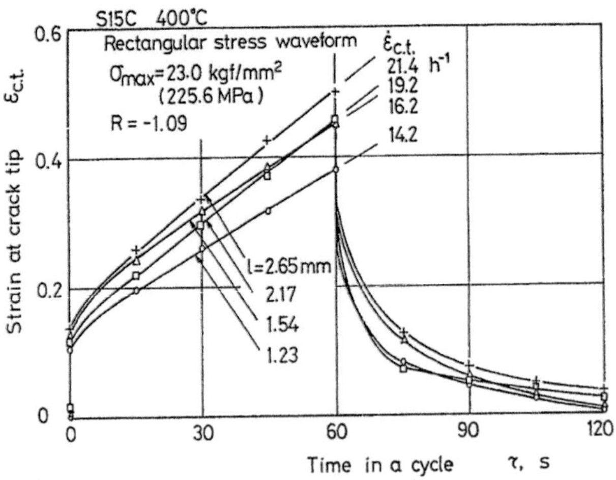
S15C 400°C
Rectangular stress waveform
σmax=23.0 kgf/mm²
(225.6 MPa)
R = -1.09
ε̇c.t.
21.4 h⁻¹
19.2
16.2
14.2
l=2.65mm
2.17
1.54
1.23
Strain at crack tip εc.t.
Time in a cycle τ, s

(b)

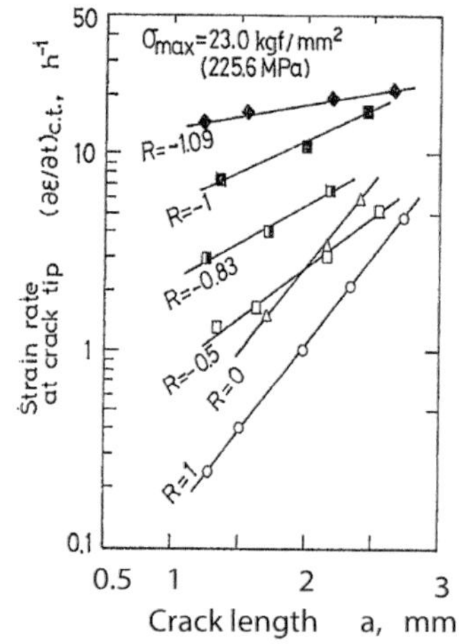

(c)

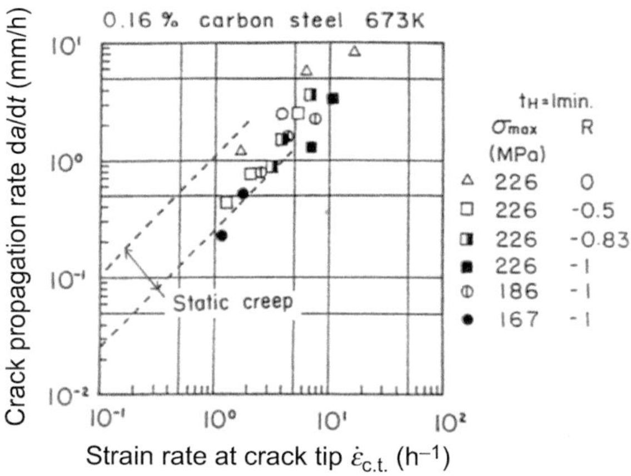

(d)

Figure 6.3 Strain near the crack tip in cyclic creep of 0.16% carbon steel at 673 K with different stress ratio [Taira *et al.*, 1980]. (a) Effect of compressive stress on the crack propagation rate, d*a*/d*N*. (b) Change in strain near the crack tip, ε_{ct}, during tensile stress hold period. (c) Effect of compressive stress on strain rate near the crack tip at tensile stress hold. (d) Relationship between the crack propagation rate, d*a*/d*t*, and the strain rate at the crack tip, $\dot{\varepsilon}_{ct}$ [Kitamura, 1986].

6.3 Acceleration due to Transition from Small-Scale Creep to Extensive Creep

In cyclic creep, if linear elastic strain predominates at the reloading instance, the transition from small-scale creep to large-scale creep occurs in each cycle of time-dependent fatigue. However, few successful experimental results have been reported on crack propagation under such a transition attributed to the mechanical factor until now.

Some results claim to include the creep-dominated crack propagation influenced by the transition. Figure 6.4 shows the results of crack propagation in cyclic creep of Inconel 718 at 923 K [Kuwabara *et al.*, 1988], encompassing various stages of the

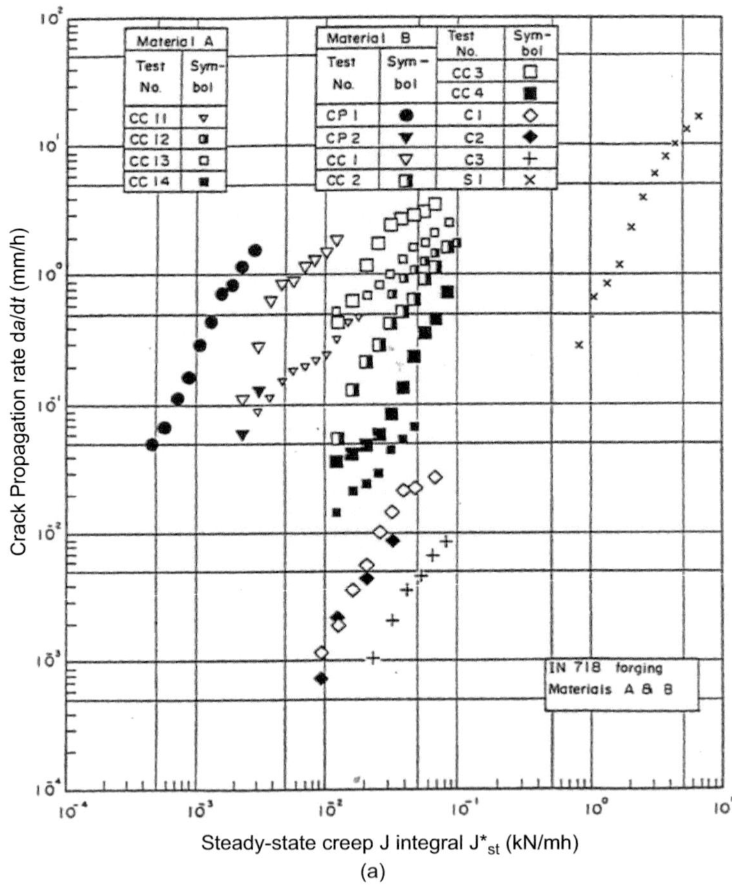

(a)

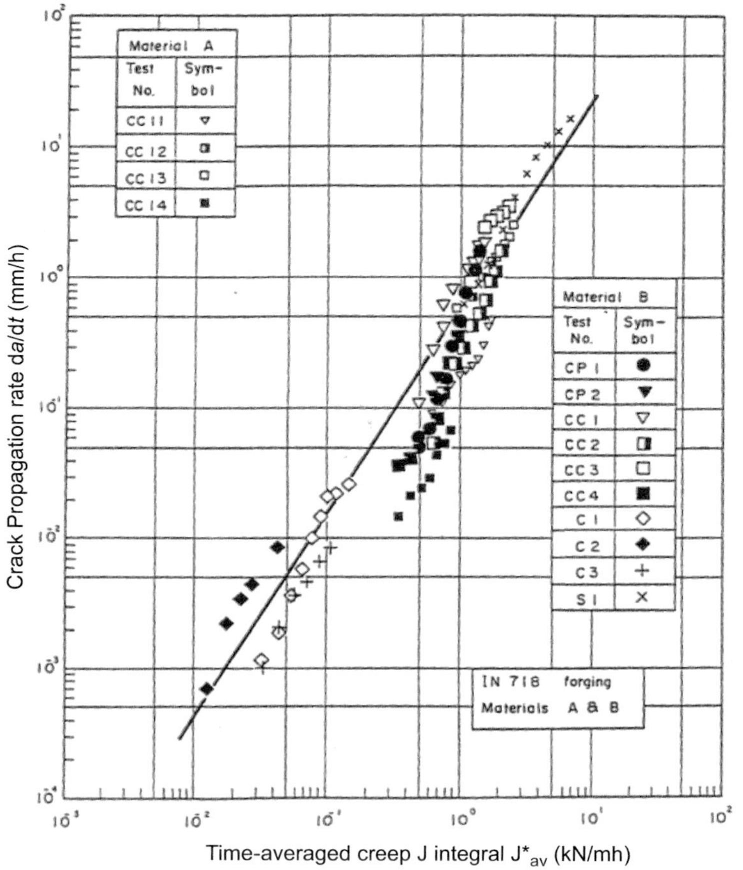

(b)

Figure 6.4 Crack propagation under the transition from small-scale creep to large-scale creep in cyclic creep of Inconel 718 [Kuwabara *et al.*, 1988]. (a) Relationship between the crack propagation rate, $(da/dt)_{av}$, and the creep J-integral evaluated based on steady-state creep, J^*_{st}. (b) Relationship between the crack propagation rate, $(da/dt)_{av}$, and the creep J-integral evaluated based on the transition from small-scale creep to large-scale creep.

transition by widely varying the tensile stress holding times t_H. Figure 6.4(a) plots the averaged crack propagation rate, $(da/dt)_{av}$, against the steady-state creep J-integral J^*_{st}, evaluated by assuming the extended creep state (large-scale creep). The results of C1, C2, and C3 with long tensile stress hold time exhibit a relatively good relationship with the results of static creep (S1), indicating crack propagation governed by the large-scale creep condition.

Other experiments with the shorter tensile hold time demonstrate significant acceleration on $(da/dt)_{av}$. Furthermore, they do not show good correlation with the linear elastic stress intensity factor, *K*.

Figure 6.4(b) shows the relationship between $(da/dt)_{av}$ and the averaged creep J-integral, $(J^*)_{av}$, taking into account the influence of transition (Eq. (4.6)). Including the experimental conditions with short tensile holding times, $(da/dt)_{av}$, is correlated well with $(J^*)_{av}$.

It should be noted that experimental proof is limited for this transition, and detailed features remain for future work.

As shown in Section 4.3.2, the creep-dominated region of CFRP (Carbon Fiber Reinforced Plastic) is strongly constrained in high temperature creep, resulting in crack propagation in small-scale creep state (Fig. 4.11). In cyclic creep of CFRP, $(da/dt)_{av}$ is governed by the energy release rate (or the linear elastic stress intensity factor, *K*) as presented in Fig. 6.5 [Ohtani *et al.*, 1993].

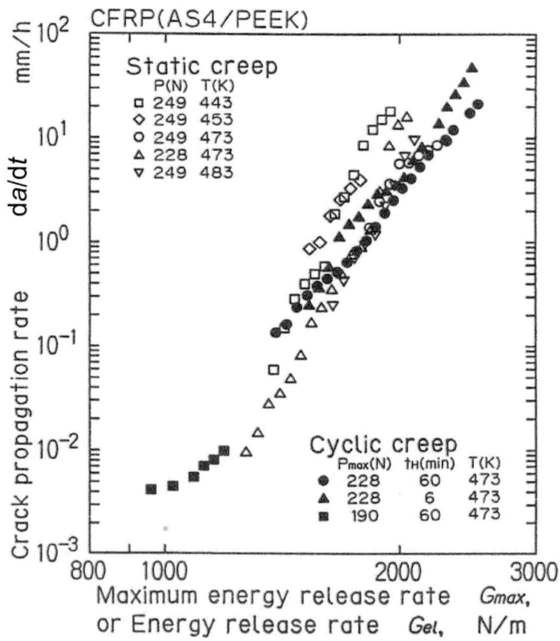

Figure 6.5 Relationship between the crack propagation rate, $(da/dt)_{av}$, and the energy release rate, G_{el}, in cyclic creep of a carbon-fiber-reinforced thermoplastic polymer at 473 K [Ohtani *et al.*, 1993].

6.4 Influence of Various Factors on Crack Propagation in Time-Dependent (Creep-Dominated) Fatigue

6.4.1 Crack Propagation in Fatigue with Stress Decreasing (Relaxation)

Figure 6.6 shows the experimental results of crack propagation under relaxation of 0.16% carbon steel at 673 K with significant

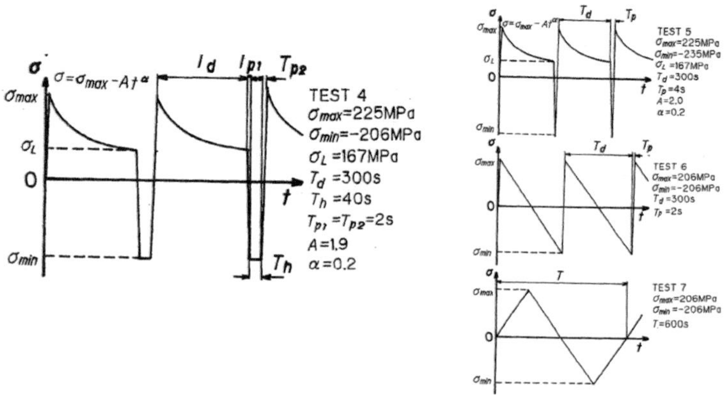

(a)

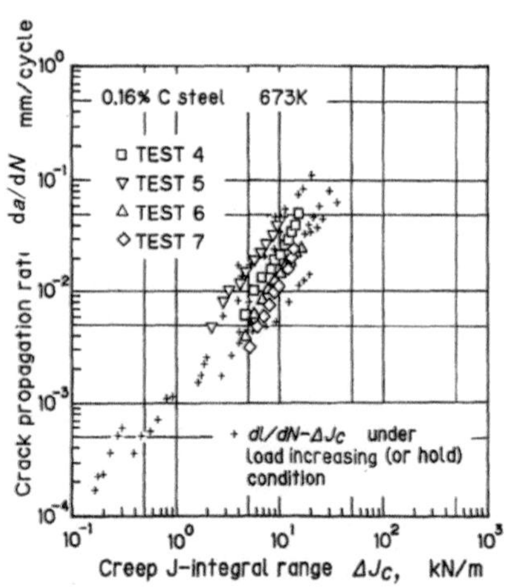

(b)

Figure 6.6 Crack propagation in cyclic creep with stress-decrease of 0.16% carbon steel at 673 K [Ohtani *et al.*, 1988]. (a) Stress conditions applied. (b) Relationship between the crack propagation rate, d*a*/d*N*, and the creep J-integral range, ΔJ_c.

stress change in every cycle [Ohtani *et al.*, 1988]. da/dN exhibits a good relationship with ΔJ_c. Additionally, the da/dN–ΔJ_c relationship is consistent with that in static creep. Here, ΔJ_c under relaxation is evaluated using the crack center opening displacement during stress reduction based on the method explained in Section 7.5.

6.4.2 Influence of Temperature

Figure 6.7 shows the dependence of the crack propagation rate and the averaged creep J-integral on temperature in time-dependent (creep-dominated) fatigue of 304 stainless steel under the same conditions except for temperature [Kitamura, 1986]. The strong exponential dependence of J^* clearly reflects the deformation due to the thermal activation process. Similarly, the dependence of da/dt indicates that crack propagation is governed by the same activation process. Thus, the da/dt–J^* relationship remains independent of temperature.

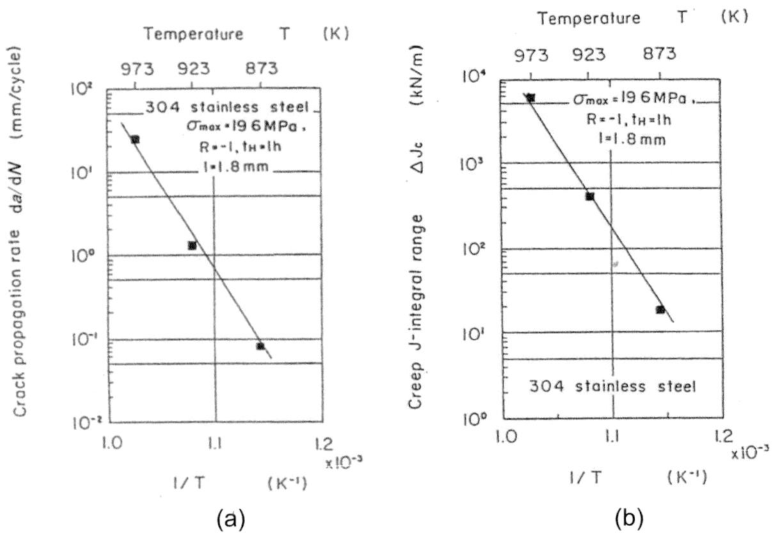

Figure 6.7 Dependence of crack propagation in cyclic creep of 304 stainless steel on temperature [Kitamura, 1986]. (a) The crack propagation rate da/dN. (b) The creep J-integral range.

6.4.3 Influence of Compressive Creep

It has been pointed out that there exists a strain (stress) waveform effect in creep-dominated fatigue conditions [Manson, 1973; Halford *et al.*, 1973]. Figure 6.8 shows the relationships between da/dN and ΔJ_c ($J^* = \Delta J_c/t_H$ where t_H = 300 seconds) in time-dependent (creep-dominated) fatigue of 304 stainless steel at 923 K without compression creep (indicated as "cp" in the figure) and with it (indicated as "cc"), respectively. The results indicate that there is no eminent waveform effect [Ohtani *et al.*, 1982].

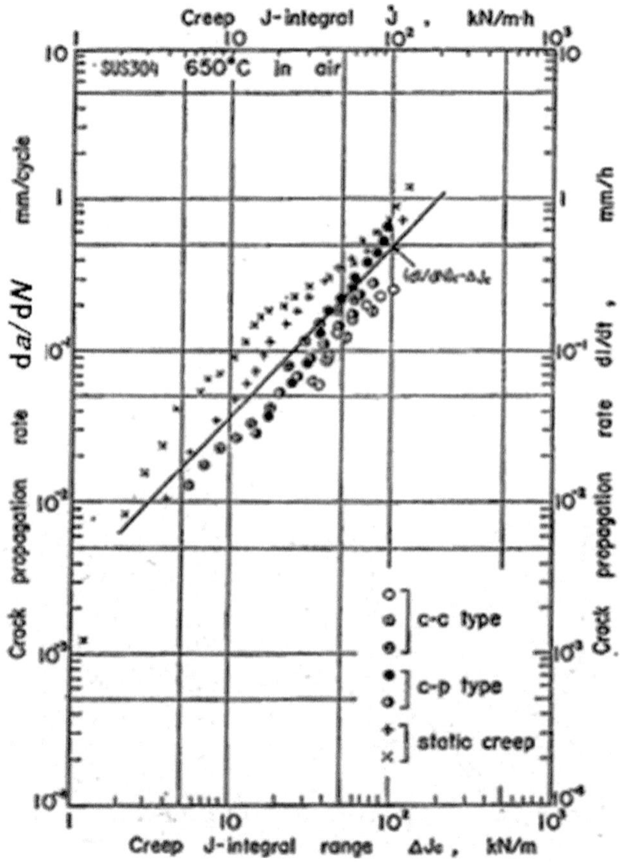

Figure 6.8 Influence of compressive creep on the relationship between the crack propagation rate, da/dN, and the creep J-integral range, ΔJ_c in cyclic creep of 304 stainless steel at 923 K [Ohtani *et al.*, 1982]. "cp" presents the cyclic creep without compressive creep while "cc" does the one with it.

6.4.4 Time-Dependent Crack Propagation of Various Steels and Alloys

Figure 6.9 [Ohtani and Kitamura, 1988] presents a summary of crack propagation obtained in the author's laboratory in time-dependent fatigue of various materials, demonstrating that the relationships fall within a narrow band.

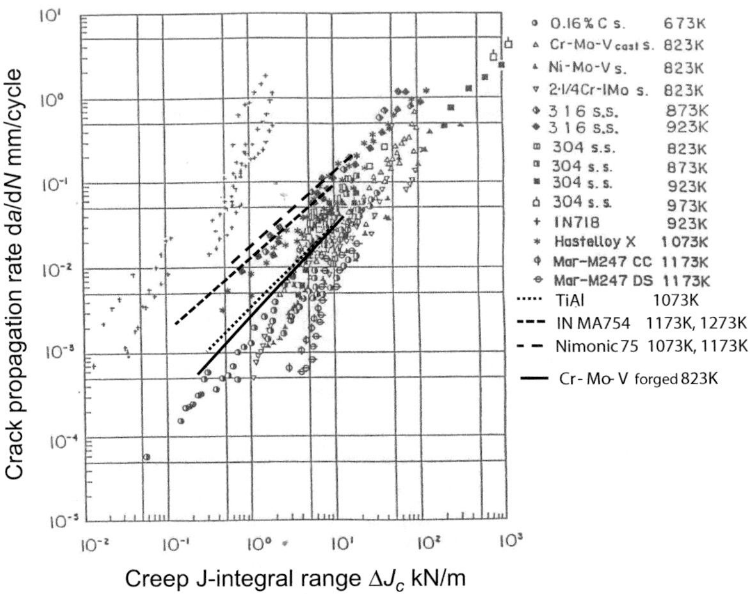

Figure 6.9 Crack propagation properties in time-dependent fatigue of various alloys and steels [Kitamura, 1986].

References

Halford, G.R., M.H. Hirschberg, and S.S. Manson, 1973, Temperature effects on the strainrange partitioning approach for creep fatigue analysis, ASTM STP, pp. 658–669.

Kitamura, T., 1986, Fracture Mechanics Investigation on Crack Propagation in High Temperature Fatigue, Doctoral Thesis.

Kitamura T., H.Sugihara, and R. Ohtani, 1986, Effect of cyclic stress changes on creep crack propagation, *Journal of The Society of Materials Science Japan*, 35, pp. 260–266. https://www.jstage.jst.go.jp/article/jsms1963/35/390/35_390_260/_pdf/-char/ja

Kuwabara K., A. Nitta, T. Kitamura, and T. Ogata, 1988, Effect of small-scale creep on crack initiation and propagation under cyclic loading, ASTM STP 924, pp. 41–59.

Manson S.S., 1973, The challenge to unify treatment high temperature fatigue—A partisan proposal based on strainrange partitioning, ASTM STP 520, pp. 744–782.

Ohji, K., K. Ogura, and S. Kubo, 1978, Simplified evaluation method of J-integral under gross plasticity and application to creep cracking (in Japanese), *Transaction of Japanese Society of Mechanical Engineers*, 44, pp. 1831–1838. https://www.jstage.jst.go.jp/article/kikai1938/44/382/44_382_1831/_pdf/-char/ja

Ohtani, R., and T. Kitamura, 1988, Characterization of high temperature strength of metals based on the mechanics of crack propagation, In *High Temperature Creep-Fatigue*, R. Ohtani, M. Ohnami and T. Inoue (eds.), Elsevier Applied Science, London, 65–90.

Ohtani, R., K. Yamada, T. Kashiwagi, and H. Matsubara, 1982, Crack propagation of high-temperature fatigue in 304 stainless steel, *Transaction of Japanese Society of Mechanical Engineers*, 48, pp. 1378–1390. https://www.jstage.jst.go.jp/article/kikaia1979/48/435/48_435_1378/_pdf/-char/ja

Ohtani, R., T. Kitamura, and S. Enomoto, 1988, Creep crack propagation under decreasing load condition (Japanese), *Journal of the Society of Materials Science Japan*, 37, pp. 283–288. https://www.jstage.jst.go.jp/article/jsms1963/37/414/37_414_283/_pdf/-char/ja

Ohtani R., T. Kitamura, Y. Uematsu, and M. Hojo, 1993, Mode I delamination of carbon fiber reinforced thermoplastic polymer under static and cyclic creep at elevated temperatures, ASTM STP 1174, pp. 23–38.

Taira, S., R. Ohtani, T. Kitamura, and K. Yamada, 1979, J-integral approach to crack propagation under combined creep and fatigue condition (in Japanese), *Journal of The Society of Materials Science Japan*, 28, pp. 414–420. https://www.jstage.jst.go.jp/article/jsms1963/28/308/28_308_414/_pdf/-char/ja

Taira, S., R. Ohtani, T. Yonekura, M. Osada, and T. Kitamura 1980, Time-dependent fatigue crack propagation at high temperatures (in Japanese), *Transaction of Japanese Society of Mechanical Engineers A*, 46, pp. 861–869. https://www.jstage.jst.go.jp/article/kikaia1979/46/408/46_408_861/_pdf/-char/ja

Yoon, K.B., A. Saxena, and P.K. Liaw, 1992, Characterization of creep-fatigue crack growth behavior under trapezoidal waveshape using C_t-parameter, *International Journal of Fracture*, 59, pp. 95–114.

Chapter 7

Crack Propagation in Creep–Fatigue Interaction: Fracture Mechanics Laws in High Temperature Fatigue

7.1 Interaction Analysis from Perspective of J-Integrals

Figure 7.1(a) shows experimental results in fatigue of Cr-Mo-V forged steel at 823 K under identical conditions except the cycle period, ν, corresponding to the schematic diagram depicted in Fig. 5.1(c). Three regions ((1)–(3) in Section 5.1) are identified, and the focus of this chapter is placed on the feature of region (3), medium cycle period region (10^1 seconds < $1/\nu$ < 5×10^2 seconds). Since the experiments are conducted using a trapezoidal stress waveform as illustrated in Fig. 7.1(b), elastoplastic strain takes place during rapid loading process arising of only the fatigue J-integral range, ΔJ_f, while creep strain does during stress hold period resulting in emergence of only the creep J-integral range, ΔJ_C. This indicates that the J-integral ranges can be separately evaluated, and a set of (ΔJ_f, ΔJ_C) is obtained in each cycle. The experiments are conducted not only under the conditions shown in Fig. 7.1(a) but also under different stress levels, hold times, and crack lengths, providing many sets of (ΔJ_f, ΔJ_C). By extracting the data for ΔJ_f = 10 kN/m, da/dN can be plotted against

High Temperature Fracture Mechanics
Takayuki Kitamura
Copyright © 2025 Jenny Stanford Publishing Pte. Ltd.
ISBN 978-981-5129-55-7 (Hardcover), 978-1-003-65328-8 (eBook)
www.jennystanford.com

ΔJ_c under a constant magnitude of ΔJ_f, as shown in Fig. 7.1(c). da/dN is proportional if creep governs crack propagation, while da/dN is constant if fatigue dominates it. Thus, this plot can elucidate the mixture of creep and fatigue in terms of the fracture mechanics parameter, which represents the stress/strain singular fields near the crack tip. Surprisingly, the plot clearly shows two distinctly divided regions and no medium region, indicating no creep–fatigue interaction. The discrepancy of regions between Figs. 7.1(a) and (c) can be explained as below [Kitamura and Ohtani, 1986].

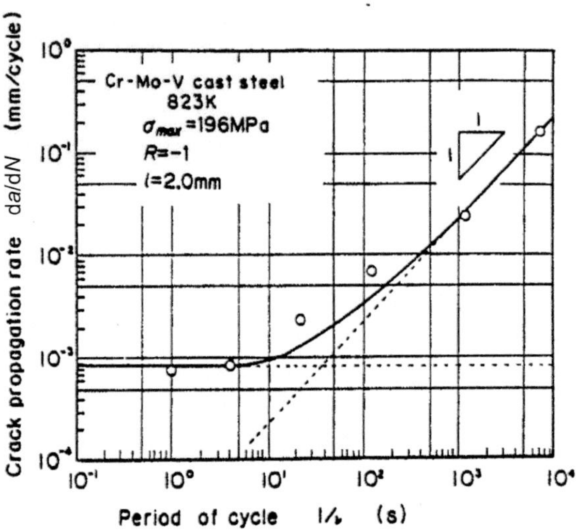

(a)

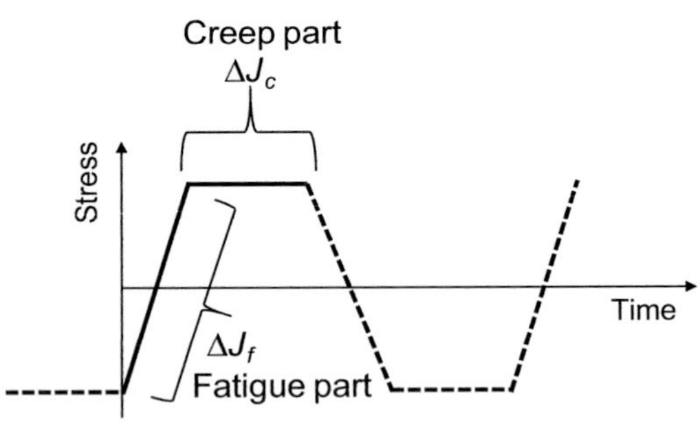

(b)

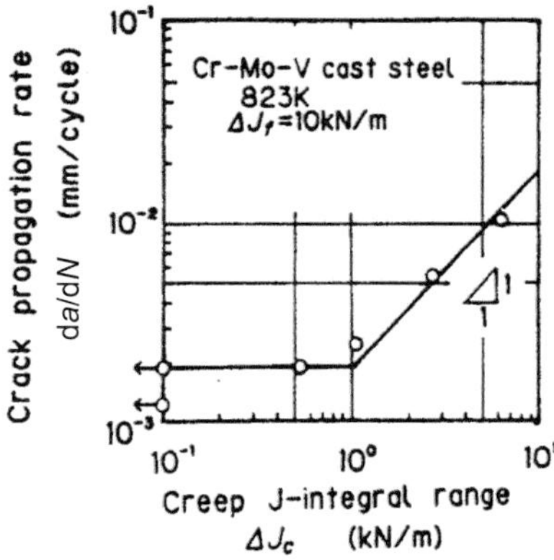

(c)

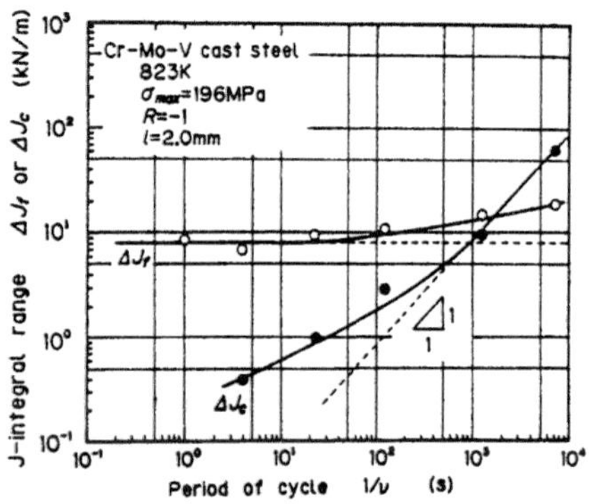

(d)

Figure 7.1 Crack propagation under interaction region in cyclic creep of Cr-Mo-V forged steel at 823 K [Kitamura and Ohtani, 1986]. (a) Relationship between the crack propagation rate, da/dN, and the stress-cycle period 1/ν. (b) Schematic illustration of evaluation of the J-integrals (ΔJ_f, ΔJ_C) in a cycle of cyclic creep. (c) Relationship between the crack propagation rate, da/dN, and the creep J-integral, ΔJ_C, under the constant magnitude of the fatigue J-integral, ΔJ_f = 10 kN/m, (d) Dependence of the J-integrals, ΔJ_f and ΔJ_C, on stress-cycle period, 1/ν.

When the stress hold time is sufficiently long, the portion of steady-state creep is dominant, and da/dN is proportional to the period of cycle (constant da/dt). In the medium region (3) of Fig. 7.1(a), ΔJ_C is not proportional to the cycle period (10^1 seconds < $1/\nu$ < 5×10^2 seconds) as presented in Fig. 7.1(d), meaning non-constant J^*. Since da/dN in transient creep is solely governed by ΔJ_C

as explained in Chapter 6, the influence of transient creep is absorbed in the $da/dN–\Delta J_C$ relationship. In other words, the discrepancy of the medium region (3) is attributed to the strong influence of transient creep (interaction of fracture driving force). It should be noted that there is no creep–fatigue interaction from perspective of fracture resistance (C_f, m_f, and C_c in Eqs. (5.2) and (5.4)).

Thus, crack propagation behavior is summarized as

(1) $\Delta J_C < 1$ kN/m
cycle-dependent crack propagation where da/dN is governed by ΔJ_f

(2) $\Delta J_C > 1$ kN/m
time-dependent crack propagation where da/dN is governed by ΔJ_C

7.2 Resistance for Cycle-Dependent and Time-Dependent Fracture

The previous section points out that there are only two main mechanisms in high temperature fatigue. Since the dimensions of ΔJ_f and ΔJ_c are the same, the $da/dN–\Delta J_f$ and $da/dN–\Delta J_C$ relationships can be plotted on the same diagram [Taira *et al.*, 1979a], as schematically illustrated in Fig. 7.2. For the same magnitude of J-integral range, da/dN in time-dependent crack propagation is faster than that in cycle-dependent one. Additionally, the latter has a slightly steeper slope than the former. This is a general feature in high temperature fatigue of various steels and alloys, including superalloys [Ohtani and Kitamura, 1988]. Considering that the J-integral range represents the strain appeared near the crack tip during one cycle (fracture driving force), this indicates that creep strain has a significantly stronger influence on local fracture at the crack tip (the amount of crack propagation) compared to elastoplastic strain, indicating a significant difference in the fracture resistance stemmed from the fracture mechanism. Differences are sometimes observed in the morphology of fracture surface and/or crack, such as the difference between inter-granular fracture and trans-granular one, or the difference between cavities and striations, but they cannot always be clearly observed.

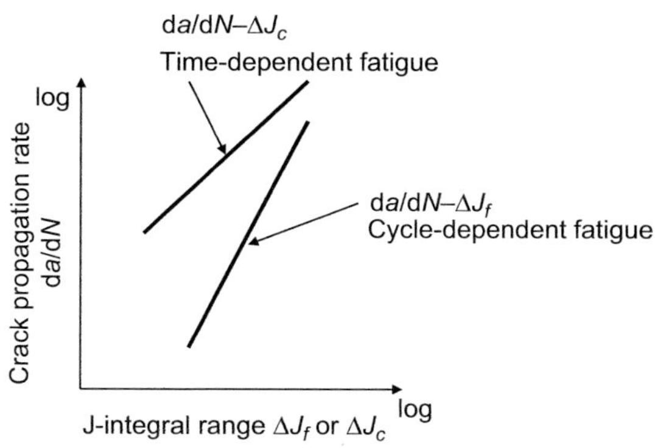

Figure 7.2 Schematic diagram of the crack propagation laws in high temperature fatigue. Relationships between the crack propagation rate, da/dN, and the J-integral ranges, ΔJ_f and ΔJ_c in cycle-dependent fatigue and time-dependent fatigue, respectively.

7.3 Crack Propagation Laws in High Temperature Fatigue

7.3.1 Dividing Law

To generalize the results in Section 7.1, equating

$$\frac{da}{dN} = \Delta J_f^{\;m_f} \qquad (7.1) = (5.2)$$

and

$$\frac{da}{dN} = C_c \Delta J_c \qquad (7.2) = (6.6)$$

the dividing law that separate cycle-dependent propagation and time-dependent propagation can be introduced as

$$\Delta J_c = \frac{C_f}{C_c} \Delta J_f^{\;m_f} \qquad (7.3)$$

Crack propagation is cycle-dependent if the left in Eq. (7.3) is less than the right, and it is time-dependent if the left is greater than the right [Kitamura and Ohtani, 1986].

Although cycle-dependent fatigue exhibits a threshold, as pointed out in Section 5.2.1 (Chapter 5), it is reported a crack propagates below the threshold [Ohji *et al.*, 1985]. This phenomenon occurs in fatigue under vibration stress with $R > 0$ (R: stress ratio), where crack propagation is driven by creep, a type of dynamic creep explained in Section 4.1.1 (Chapter 4), resulting in a good correlation between da/dt and J^*. Therefore, the dividing condition described above has limitation

$$\Delta J_f \geq \Delta J_{f\,th} \tag{7.4}$$

7.3.2 Combination Rule of Crack Propagation in Creep Fatigue

(A) Independent crack propagation rule

The results shown in Section 7.2 indicate that there is no interaction between cycle-dependent and time-dependent crack propagation. In other words, they do not coexist within a cycle, and the dominant one governs the propagation. Observations of fracture surfaces often support this, as the surface is occupied by fractures resulting from one dominant mechanism. Additionally, the observation of crack propagation within a cycle, as explained in Section 8.3 (Chapter 8), also suggests no interaction [Kitamura and Ohtani, 1986]. This implies that cycle-dependence and time-dependence are independent of each other.

Fracture Mechanics Law in High Temperature Fatigue under Independent Rule

In the case of $\Delta J_c < \dfrac{C_f}{C_c} \Delta J_f^{\,m_f}$

$$\frac{da}{dN} = C_f \Delta J_f^{\,m_f}$$

In the case of $\Delta J_c > \dfrac{C_f}{C_c} \Delta J_f^{\,m_f}$

$$\frac{da}{dN} = C_c \Delta J_c$$

(B) Linear summation crack propagation rule

The linear damage accumulation concept has been historically adopted to analyze the fracture process and mechanism in creep–fatigue interaction at high temperatures. Applying to crack propagation within a cycle, we can introduce another reasonable rule—the crack extension in a cycle is given by the sum of cycle-dependent and time-dependent crack propagation [Saxena, 1993]. These, of course, are governed by ΔJ_f and ΔJc as shown in Eqs. (7.1) and (7.2), respectively. An advantage is no need for the dividing condition.

Fracture Mechanics Law in High Temperature Fatigue under Summation Rule

$$\frac{da}{dN} = C_f \Delta J_f^{\ m_f} + C_c \Delta J_c$$

Although the difference between (A) Independent rule and (B) Linear summation rule may seem significant, the difference in d*a*/d*N* is only at most twice. Considering the dispersion of experimental results due to implementation difficulties and the ambiguity of usage conditions in actual equipment, this is not practically a severe issue from an industrial perspective [Taira *et al.*, 1979b]. Furthermore, the linear summation rule tends to provide conservative prediction and has a simple expression.

7.3.3 Comprehensive Understanding of Phenomena at High Temperatures

Strength at high temperatures have been referred to by various names under different load/temperature conditions. The laws of crack propagation in the previous sections provide comprehensive understanding of such phenomena in terms of the fracture mechanics.

For an example, "dynamic creep" and "cyclic creep" fundamentally possess the same fracture mechanism/mechanics at the crack tip as "static creep," resulting in universal creep crack propagation.

As another example, the difference between "dynamic creep" and "high-cycle fatigue" can be clearly explained by the dividing law, Eq. (7.3). The fraction of creep (time-dependent) and elastoplastic (time-independent) deformation governs the fracture at the crack tip and determines the time-dependent crack extension in "dynamic creep" or the cycle-dependent one in "high-cycle fatigue."

7.4 Crack Propagation in Thermal Fatigue

Temperature changes and their distribution within a component are crucial factors to consider in the fracture process of equipment exposed to high temperatures in fatigue conditions. This is referred to as thermal fatigue, resulting from thermal stress and/or mechanical stress under cyclic temperature. The applicability of the above fracture mechanics laws to thermal fatigue is one of the most important issues caused by creep–fatigue interaction in industrial applications.

As demonstrated previously, the laws exhibit weak temperature dependence as long as the fracture mechanism of creep or fatigue does not change. This suggests that the above fracture mechanics law generally holds true even in thermal fatigue. However, there are not many experimental proofs on crack propagation in thermal fatigue due to the difficulty of experimentation and exact analysis.

Figure 7.3 shows the crack propagation behavior in thermal fatigue of Cr-Mo-V steel under cyclic strain-temperature controlled conditions [Nitta *et al.*, 1989]. The experiments are conducted in two types of fatigue within a temperature range of 573 K – 823 K as schematically illustrated in Figure 7.3(a)—in-phase thermal fatigue (upper illustration) where strain and temperature change in sync, and out-of-phase thermal fatigue (lower illustration) where they change out of sync. Figure 7.3(b) depicts two distinct types of crack propagation—the da/dN–ΔJ_f relationship in cycle-dependent crack propagation and the da/dN–ΔJ_C in time-dependent (creep-dominated) one, proving that the fracture mechanics laws hold true in both types of thermal fatigue if the J-integral ranges are properly evaluated. Furthermore, they correspond well with those in isothermal fatigue. While the out-of-phase (in-phase) tends to result in cycle-dependent (time-dependent) fatigue, it is determined by the set of (ΔJ_f, ΔJ_C). Similar results are obtained for 304 stainless steel [Nitta *et al.*, 1987].

Additionally, the crack propagation characteristics are investigated under a combination of triangular temperature (673 K – 773 K) waveform and rectangular stress waveform of 0.16% carbon steel [Ohtani *et al.*, 1987].

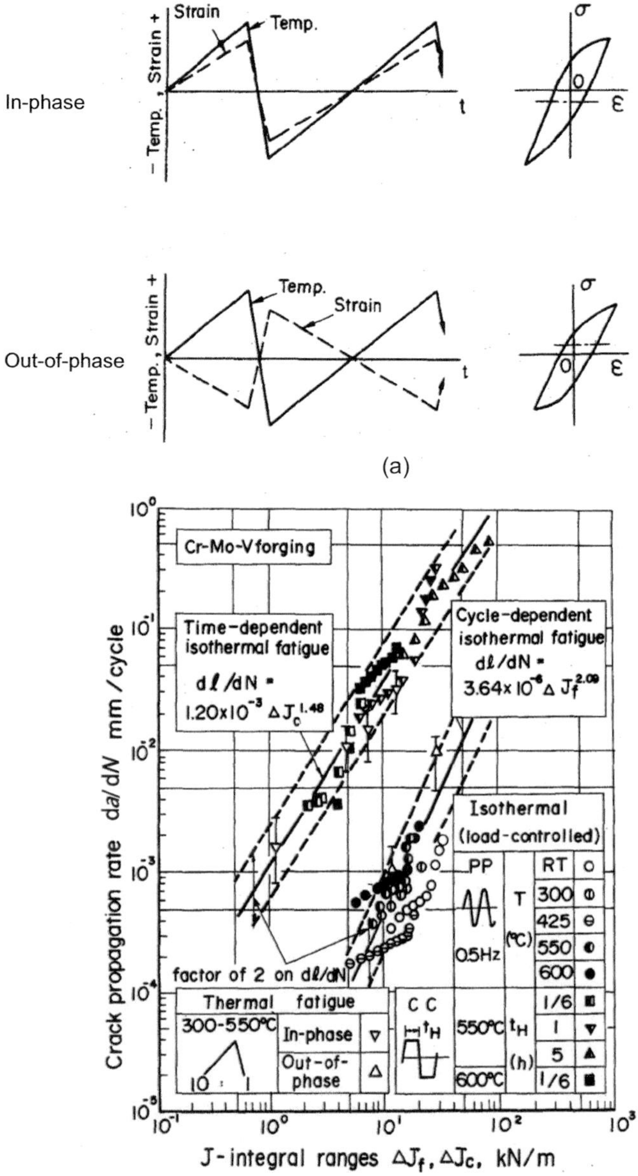

(a)

(b)

Figure 7.3 Crack propagation in thermal fatigue of Cr-Mo-V steel under 573 K – 823 K [Nitta *et al.*, 1989]. (a) Strain and temperature waveforms in thermal fatigue. (b) Relationship between the crack propagation rate, d*a*/d*N*, and the J-integral ranges, ΔJ_f or ΔJ_c, in thermal and isothermal fatigue.

It should be noted that the mechanism of deformation and fracture may vary depending on the combination of temperature and stress levels. There is a report that the fracture characteristic changes under significantly high temperature, resulting in different fracture resistance, C_C, in time-dependent (creep-dominated) crack propagation [Ohtani *et al.*, 1995]. Since no experiments have been conducted in thermal fatigue over such a wide temperature range where mechanisms may differ, future studies are required. Additionally, crack propagation in thermal fatigue with steep temperature gradients is another important issue. Experimental/analytical investigations under such general thermal fatigue conditions are necessary for accurately evaluating the fracture mechanics laws. The concept of the J-integral [Kishimoto *et al.*, 1980] might be beneficial for expanding the application of the laws for such conditions.

7.5 Experimental Evaluation Method of ΔJ_C and ΔJ_f in Creep Fatigue

In pure cycle-dependent fatigue, ΔJ_f is calculated from the area in the P–δ curve (load-displacement curve) or P–V curve (load-crack center opening displacement curve) shown in Fig. 2.9, regardless of the constitutive equation (stress–strain relation). This approach is convenient and effective because P and δ are accurately and easily measured by standard fatigue testing equipment. Although this procedure cannot be directly applied to evaluate ΔJ_C, it is possible to introduce a similar graphical method based on Eq. (3.15).

7.5.1 Evaluation Method in Cyclic Creep [Taira *et al.*, 1979a]

In cyclic creep with a rectangular stress waveform as depicted in Fig. 7.4(a), time-independent elastoplastic deformation takes place only during the rapid load-increasing period, while creep deformation does only during stress hold period. Then, by simply integrating Eq. (3.15), ΔJ_C is evaluated using the area S_c in P–δ (P–V) hysteresis loop shown in Fig. 7.4(a).

$$\Delta J_c = \int_0^{t_1} J^* dt \approx \int_0^{t_1} \sigma_{net} \dot{V} dt = \frac{S_c}{h_t b} \tag{7.5}$$

Of course, ΔJ_f is evaluated using the area S_p by

$$\Delta J_f = \frac{\Delta K^2}{E} + \frac{S_p}{h_t b} \qquad (7.6) = (2.21)$$

7.5.2 Evaluation Method in Load Decreasing (Relaxation) Conditions [Ohtani *et al.*, 1988]

Creep strain increases under conditions of slow load decreasing, including stress relaxation under strain hold. The hysteresis loop is schematically illustrated in Fig. 7.4(b), where the dotted line indicates the linear elastic unloading curve obtained from a rapid-decrease in load from the maximum stress. Since creep displacement is determined by the difference between V in the continuous cycle (unloading part) and in the rapid unloading, ΔJ_c is evaluated using the area S_c in Fig. 7.4(b) by

$$\Delta J_c = \int_0^{t_1} J^* dt \approx \int_0^{t_1} \sigma_{net} \dot{V} dt = \frac{S_c}{h_t b} \tag{7.7}$$

7.5.3 Evaluation Method of General Creep Fatigue [Taira *et al.*, 1979a]

An attempt is made to expand the method to cases where both elastoplastic deformation (displacement) $V_e + V_p$ and the creep one V_c take place simultaneously.

In order to analyze the failure life of smooth materials in high temperature fatigue, a method is proposed for separating elastoplastic and creep deformation, known as the strain-partitioning method [Halford *et al.*, 1973]. In the method, one rapid straining cycle (dotted line) is inserted during continuous fatigue cycles (continuous line), as illustrated in left of Fig. 7.4(c). The displacement δ in the continuous fatigue cycle consists of elastoplastic displacement and creep one, $\delta_e + \delta_p + \delta_c$, while δ in the rapid cycle includes only $\delta_e + \delta_p$. This indicates that creep deformation can be extracted from the difference between two hysteresis loops at a given load.

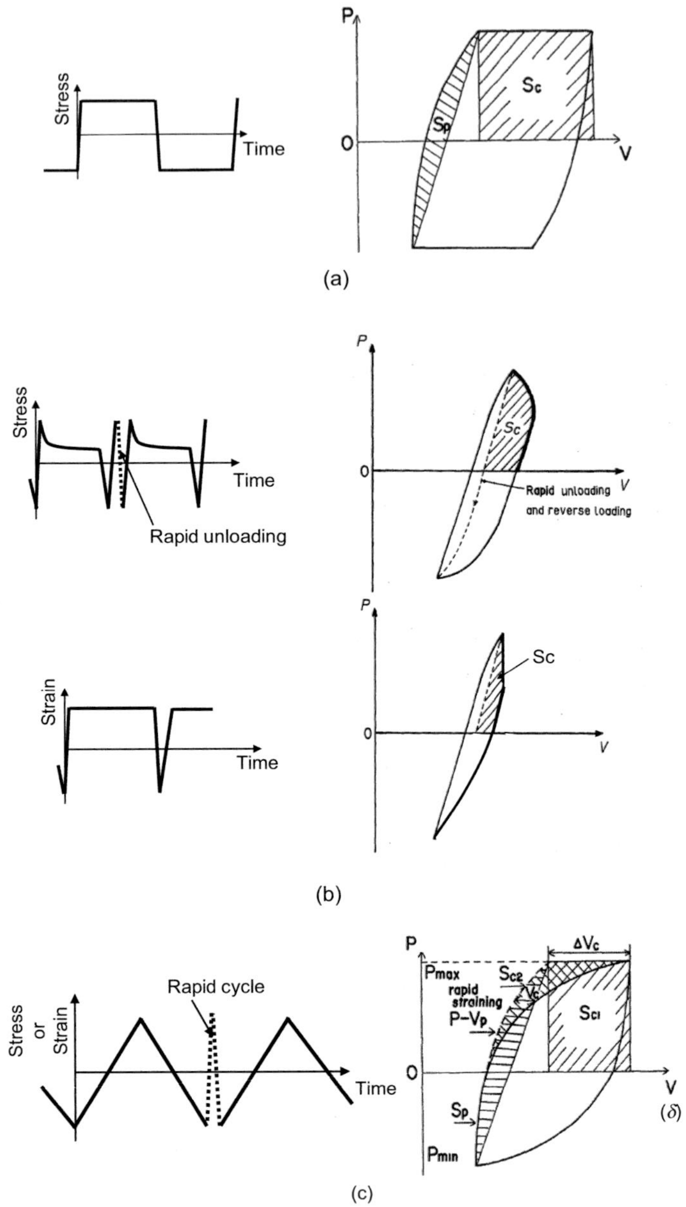

Figure 7.4 Graphic evaluation method of the J-integrals, ΔJ_f and ΔJ_C, [Taira *et al.*, 1979a; Ohtani *et al.*, 1988]. (a) Evaluation method of the J-integral ranges using P–δ (or P–V) hysteresis loop in cyclic creep. (b) Evaluation method of the creep J-integral range in fatigue cycle with decreasing creep stress. (c) Evaluation method of the J-integral ranges using P–δ (or P–V) hysteresis loop in general creep fatigue.

By applying this strain-partitioning method, the creep component in the crack opening displacement can be extracted from the loops. Fig. 7.4(c) presents the *P*–*V* hysteresis obtained from continuous fatigue cycle (continuous line) and a rapid-straining cycle (dotted line), and $V_e + V_p$ and V_c can be separated from the two hysteresis loops. ΔJ_f is evaluated from the area S_p in the hysteresis loop of rapid straining, while ΔJ_c is evaluated using the areas, S_{c1} and S_{c2}, based on the difference between the two loops as (Appendix 7-A)

$$\Delta J_c = \frac{S_{c1} - S_{c2}}{h_t b} \tag{7.8}$$

Experimental investigation indicates successful implementation of this method [Nitta *et al.*, 1989].

Since no progress has been made in analyzing the singular fields of stress and strain near the crack tip in such simultaneous elastoplastic and creep deformation, detailed analytical research on these issues is required in the future.

Appendix

7-A Evaluation of Creep J-Integral Range in General Fatigue Condition [Taira *et al.*, 1979a]

The creep component in crack opening displacement, V_c, is determined by the difference in displacement between two hysteresis loops in creep fatigue and in rapid cycle, and is assumed to be a function of load as

$$V_c = f(P) \tag{7A.1}$$

As the creep J-integral, J^*, is experimentally evaluated by

$$J^* \approx \sigma_{net} \dot{V}_c \tag{7A.2}$$

then,

$$\Delta J_c = \int_0^{t_1} J^* dt \approx \int_0^{t_1} \sigma_{net} \dot{V}_c dt = \int_0^{V_{c1}} \sigma_{net} dV_c \tag{7A.3}$$

By substituting Eq. (7A.1) into Eq. (7A.3), the following equation is derived.

$$\Delta J_c = \frac{1}{h_t b}\int_0^{P_{max}} dP = \frac{1}{h_t b}\left\{\left[Pf(P)\right]_0^{P_{max}} - \int_0^{P_{max}} f(P)dP\right\} = \frac{S_{c1} - S_{c2}}{h_t b} \tag{7A.4}$$

References

Halford, G.R., M.H. Hirschberg, and S.S. Manson, 1973, Temperature effects on the strainrange partitioning approach for creep fatigue analysis, ASTM STP, pp. 658–669.

Kishimoto, K., S. Aoki, and M. Sakata, 1980, On the Path independent integral-$\hat{J}$, *Engineering Fracture Mechanics*, 13, pp. 841–850.

Kitamura, T., and R. Ohtani, 1986, Crack propagation under interaction between creep and fatigue, (in Japanese), *Transaction of Japanese Society of Mechanical Engineers A*, 52, pp.1816–1823. https://www.jstage.jst.go.jp/article/kikaia1979/52/480/52_480_1816/_pdf/-char/ja

Manson, S.S., 1973, The challenge to unify treatment high temperature fatigue—A partisan proposal based on strainrange partitioning, ASTM STP 520, pp. 744–782.

Nitta, A., T. Ogata, and K. Kuwabara, 1989, Thermal fatigue crack propagation and failure life evaluation on Cr-Mo-V and SUS304 steels, (in Japanese), *Journal of The Society of Materials Science Japan*, 38, pp. 423–429. https://www.jstage.jst.go.jp/article/jsms1963/38/427/38_427_423/_pdf/-char/ja

Ohji, K., S. Kubo, W.S. Joo, and T. Kuri, 1985, Time-dependent and cycle-dependent crack propagation under sinusoidal pulsating stress waveform in creep temperature range of SUS304 stainless steel, (in Japanese), *Transaction of Japanese Society of Mechanical Engineers A*, 51, pp. 2137–2145. https://www.jstage.jst.go.jp/article/kikaia1979/51/469/51_469_2137/_pdf/-char/ja

Ohtani, R., and T. Kitamura, 1988, Characterization of high temperature strength of metals based on the mechanics of crack propagation, In *High Temperature Creep-Fatigue*, R. Ohtani, M. Ohnami and T. Inoue (eds.), Elsevier Applied Science, London, 65–90.

Ohtani, R., T. Kitamura, and N. Tada, 1987, Crack propagation in high temperature fatigue under cyclic temperature, (in Japanese), *Transaction of Japanese Society of Mechanical Engineers A*, 53, pp. 1030–1036. https://www.jstage.jst.go.jp/article/kikaia1979/53/490/53_490_1030/_pdf/-char/ja

Ohtani, R., T. Kitamura, and S. Enomoto, 1988, Creep crack propagation under decreasing load condition, *Journal of the Society of Materials Science Japan*, 37, pp. 283–288. https://www.jstage.jst.go.jp/article/jsms1963/37/414/37_414_283/_pdf/-char/ja

Tada N., W. Zhou, T. Kitamura, and R. Ohtani, 1995, Creep-fatigue intergranular fracture of inner cracking type in Type 304 stainless steel, in Japanese, *Journal of the Society of Materials Science Japan*, 44, pp. 84–89. https://www.jstage.jst.go.jp/article/jsms1963/44/496/44_496_84/_pdf/-char/ja

Saxena A., 1993, Fracture mechanics approaches for characterizing creep-fatigue crack growth, *JSME International Journal Ser. A*, 36, pp. 1–20.

Taira, S., R. Ohtani, T. Kitamura, and K. Yamada, 1979a, J integral approach to crack propagation under combined creep and fatigue condition (in Japanese), *Journal of The Society of Materials Science Japan*, 28, pp. 414–420. https://www.jstage.jst.go.jp/article/jsms1963/28/308/28_308_414/_pdf/-char/ja

Taira, S., R. Ohtani, and T. Komatsu, 1979b, Application of J-integral to high-temperature crack propagation, Part II Fatigue crack propagation, *Journal of Engineering Materials and Technology*, 101, pp. 162–167.

Chapter 8

Complexity Caused by Creep–Fatigue Interaction: For Future Research

8.1 Difference in Crack Propagation between Static Creep and Time-Dependent (Creep-Dominated) Fatigue

Creep-dominated fatigue involves periodic unloading or compression half-cycles. Subsequent reloading significantly accelerates not only the tensile creep strain rate near the crack tip but also the creep crack propagation rate in each cycle, resulting in a good correlation between $\mathrm{d}a/\mathrm{d}t$ and J^* or $\mathrm{d}a/\mathrm{d}N$ and ΔJ_c. However, detailed experimental observations have highlighted subtle contradictions. Figure 8.1(a) [Kitamura *et al.*, 1986] compares the $\mathrm{d}a/\mathrm{d}N$–ΔJ_c relationships between static creep and cyclic creep of 304 stainless steel at 923 K. $\mathrm{d}a/\mathrm{d}t$ and J^* in static creep is converted into $\mathrm{d}a/\mathrm{d}N$ and ΔJ_c as the value per hour indicated by the dashed line band. The results in cyclic creep, shown by square marks, are located below those in static creep. Similarly, Fig. 8.1(b) compares the $\mathrm{d}a/\mathrm{d}t$–J^* relationship in crack propagation in dynamic and static creep. The results in dynamic creep are near the lower limit of those in static

High Temperature Fracture Mechanics
Takayuki Kitamura
Copyright © 2025 Jenny Stanford Publishing Pte. Ltd.
ISBN 978-981-5129-55-7 (Hardcover), 978-1-003-65328-8 (eBook)
www.jennystanford.com

creep. Both indicate that creep crack propagation in time-dependent fatigue is slower than that of static creep for the same fracture driving force. As shown in Fig. 6.2(c), this is a general characteristic, although the difference varies depending on the cycle conditions.

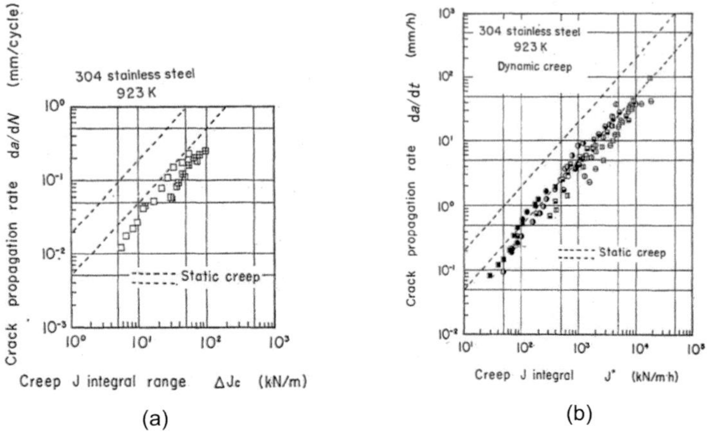

Figure 8.1 Effect of compressive stress or unloading on creep crack propagation of 304 stainless steel at 923 K [Kitamura *et al.*, 1986]. (a) Relationship between the crack propagation rate, da/dN, and the creep J-integral range, ΔJ_c in cyclic creep (square symbols) and static creep (dashed-line band). (b) Relationship between the crack propagation rate, da/dt, and the creep J-integral, J^* in dynamic creep (symbols) and static creep (dashed-line band).

Since the acceleration of strain rate due to transient creep is reflected in evaluation of J^* (fracture driving force), the trend implies that the creep strain in the reloaded tension does not bring about the same amount of fracture (fracture resistance) at the crack tip, suggesting healing of creep damage during unloading and compression. Although the mechanism is not clear at present, it is thought to be due to the disappearance of micro-cracks and cavities near the crack tip (recovery of damage) during compression, which are formed by tensile creep in the previous cycle.

Investigating experimental results under several time-dependent (creep-dominated) fatigue conditions of steels and alloys reveals influential factors contributing to differences as:

(a) Stress ratio (compressive stress), R

(b) Stress hold time in compression, t_{cH}

An empirical attempt to estimate the recovery from experimental results suggests the following equation introducing a parameter, γ:

$$\left(\Delta J_c\right)_{ef} = \frac{\Delta J_c}{\gamma} \tag{8.1}$$

where γ is defined as

$$\gamma = 1 + \frac{C_1\left(1-R\right)}{C_2 + EB\sigma_{net}^{\ \ n-1} t_H} \tag{8.2}$$

Here C_1 and C_2 are constants. Young's modulus E, B, and n in the Norton law, and the net-section stress σ_{net} are used to render the expression dimensionless. The relationship aligns with the static creep relationship, as depicted in Fig. 8.2 [Kitamura *et al.*, 1986].

As this modification is solely an empirical formula based on limited experimental results, further research is necessary to elucidate the mechanism and mechanics of damage recovery. On the other hand, it is worthy of noting that from an industrial perspective the original fracture mechanics law provides a conservative estimation even when the effect of damage recovery is ignored.

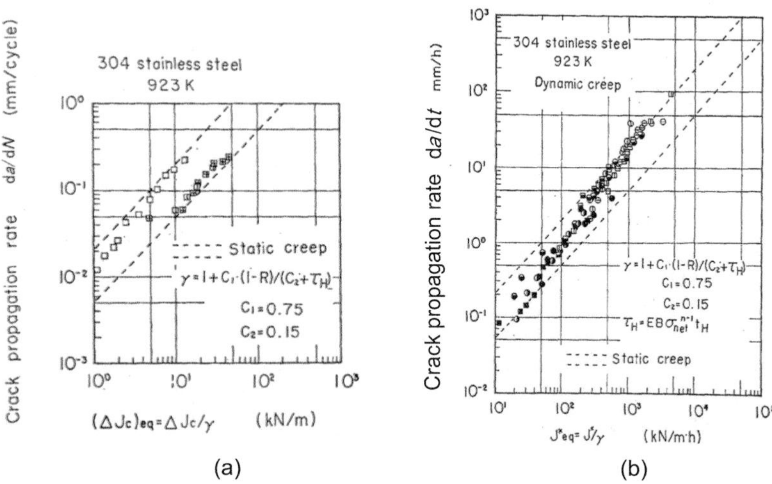

Figure 8.2 Modification of compressive stress or unloading effect [Kitamura *et al.*, 1986]. (a) Relationship between the crack propagation rate and the effective creep J-integral range in cyclic creep (square symbols) and static creep (dashed-line band). (b) Relationship between the crack propagation rate and the effective creep J-integral in dynamic creep (symbols) and static creep (dashed-line band).

8.2 Influence of Pre-damage on Crack Propagation

8.2.1 Crack Propagation of Pre-damaged Materials

In materials used in actual components, which do not initially include intrinsic defects, a main crack may form after experiencing creep and/or fatigue loads. Pre-existing damage from creep and fatigue occurring before crack initiation could influence crack propagation.

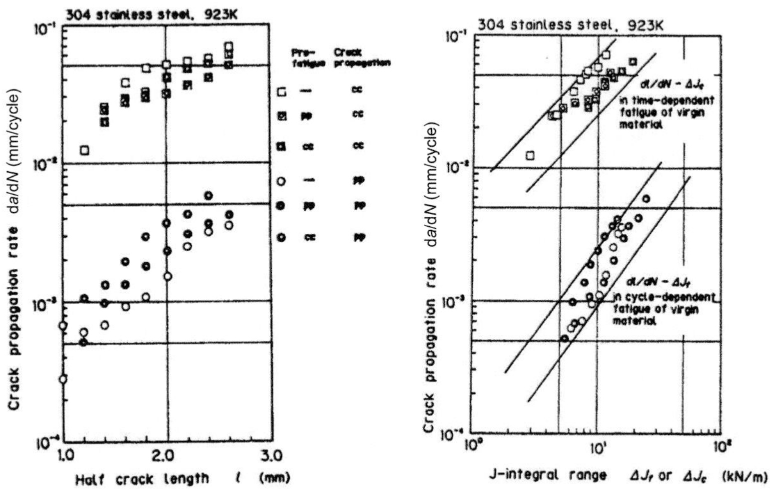

(a) (b)

Figure 8.3 Influence of pre-damage on crack propagation in cycle-dependent fatigue and time-dependent fatigue. “pre-cc damage” and “pre-pp damage” mean materials that the damage is introduced by creep-dominated fatigue and cycle-dependent fatigue, respectively, before the crack propagation tests [Kitamura and Ohtani, 1986]. (a) Relationship between the crack propagation rate, d*a*/d*N*, and the crack length, $l = a$. (b) Relationship between the crack propagation rate, d*a*/d*N*, and the fatigue J-integral range, ΔJ_f or the creep J-integral range, ΔJ_c.

Figure 8.3 [Kitamura and Ohtani, 1986] shows the crack propagation behavior in cycle-dependent fatigue (pp) and time-dependent (creep-dominated fatigue (cc) of pre-damaged 304 stainless steel at 923 K. The pre-damage is introduced in a smooth specimen before forming a starter crack through cycle-dependent fatigue (pre-pp) or time-dependent fatigue (pre-cc) at 923 K until

85% of its rupture life. Even in creep-dominated fatigue, small cracks are initiated on only the surface of smooth specimen.

Figure 8.3(a) demonstrates that pre-damage has minimal influence on da/dN compared to results from materials without pre-damage, resulting in the same da/dN $-\Delta J_f$ or da/d$N$$-\Delta J_c$ relationship as shown in Fig. 8.3(b). Previously introduced damage, small cracks on the surface, influences not only elastoplastic or creep deformation (fracture driving force) but also fracture characteristics (fracture resistance) at the crack tip. It should be noted that while deformation characteristics may be affected by pre-damage (previous stress cycles) in some cases, both the crack propagation rate and the fatigue or creep J-integral range change simultaneously, leading to unchanged fracture mechanics laws.

Nevertheless, this provides a warning against overly simplistic modeling assumption for the accumulation damage regarding crack propagation.

8.2.2 Influence of Severe Pre-creep Damage (Inner Small Cracks)

In time-dependent (creep-dominated) fatigue under low stress without compressive creep (cp-type), a huge number of small cracks and cavities appear throughout the interior of material [Kawasaki *et al.*, 1992], while numerous small cracks are only initiated on the surface in ordinary creep–fatigue experiments. A similar fracture process also takes place at very high temperatures and low stresses. These bring about volumetric damage, which is different from damage limited to the specimen surface.

A crack propagation experiment is conducted at 923 K in cycle-dependent fatigue of pre-damaged 304 stainless steel with an extremely high density of microscopic internal defects, which is introduced until nearly the end of life under creep-dominated fatigue at 1073 K (low stress cp-type loading). Figure 8.4(a) compares crack propagation of 304 stainless steel with pre-creep damage and without it [Ohtani *et al.*, 1995], revealing a significant acceleration of da/dN against ΔJ_f by pre-damage. While ΔJ_f does not change significantly under this condition, da/dN is greatly accelerated by pre-damage. indicating that volumetric damage mainly reduces the fracture resistance.

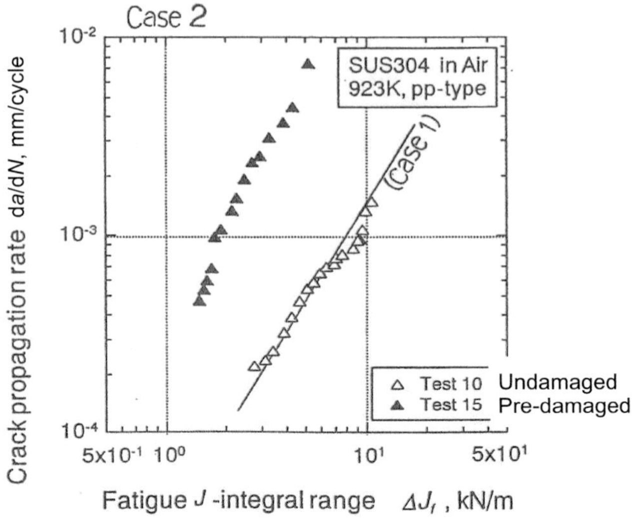

(a)

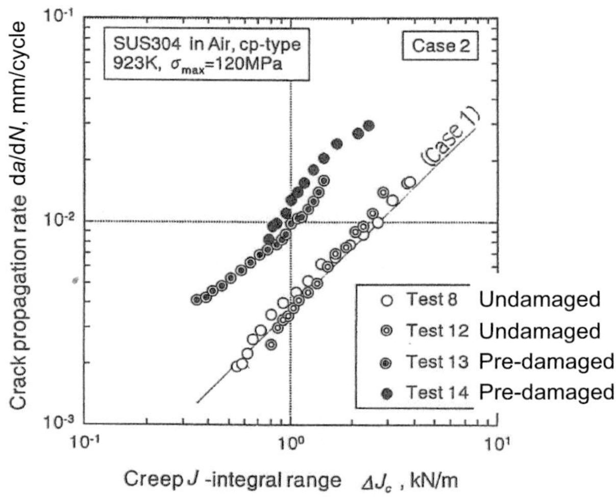

(b)

Figure 8.4 Influence of numerous inner small cracks (severe pre-damage) on subsequent crack propagation [Ohtani *et al.*, 1995]. (a) Relationship between the crack propagation rate, d*a*/d*N*, and the fatigue J-integral range, ΔJ_f, in subsequent cycle-dependent fatigue—Test 15: Damaged material, and Test 10: Non-damaged material. (b) Relationship between the crack propagation rate, d*a*/d*N*, and the creep J-integral range, ΔJ_c, in subsequent time-dependent (creep-dominated) fatigue [Ohtani *et al.*, 1995]—Tests 13 and 14: Damaged material, and Tests 8 and 12: Non-damaged material.

Similarly, severe volumetric damage also accelerates the crack propagation rate in time-dependent fatigue, as shown in Fig. 8.4(b) [Ohtani *et al.*, 1995]. In Section 3.3.4 (Chapter 3), the same influence presents the acceleration of da/dt in subsequent static creep of pre-damaged material.

Although it is evident that volumetric damage drastically accelerates subsequent crack propagation, detailed analysis of the mechanism related to acceleration has not been conducted, and no modeling has been developed to quantitatively predict the degree of acceleration. Although this is one of the unsolved issues for the future, it is necessary to elucidate physical substance and evolution mechanism of damage for developing the damage mechanics perspective.

8.2.3 Cracking Behavior Near Notch Root at Early Stage

It is well known that the da/dt–J^* relationship at the early stage of creep crack propagation immediately after initiation from a notch root differs from that under steady propagation. da/dt at this stage is slower for the same magnitude of J^*, often attributed to immature damage near the crack tip. It is the formation process of a stable fracture progression region, namely the damage process zone. This phenomenon also points out the complexity of the damage forming process and its influence on subsequent crack propagation. Investigation of this process may provide valuable insights into the fracture mechanism at the crack tip and the physical nature of creep damage.

8.3 Complexity of Crack Propagation during One Cycle

8.3.1 Observation of Crack Propagation during One Cycle in Cyclic Creep

Cyclic creep illustrated in Fig. 8.5(a) simply makes us image crack propagation

$$\frac{da}{dN} = \Delta a_U + \Delta a_H \tag{8.3}$$

which is the linear summation crack propagation rule in Section 7.3.2(B) (Chapter 7). Here, Δa_U and Δa_H are the amounts of crack extension during spike-like loading and stress hold, respectively. Since they are purely caused by cyclic elastoplastic and creep deformation, respectively, the former is correlated with ΔJ_f and the latter is dominated by ΔJ_c through Eqs. (7.1) and (7.2). On the other hand, the independent crack propagation rule in Section 7.3.2(A) suggests

$$\frac{da}{dN} = \Delta a_H \tag{8.4}$$

Continuous measurement of crack propagation during one cycle greatly contributes to elucidating crack propagation mechanics under creep–fatigue interaction. Such a challenge is conducted under cyclic creep of 304 stainless steel at 923 K [Kitamura and Ohtani, 1986]. Under this condition, no compressive creep is applied during one cycle, so the damage recovery explained in Section 8.1 is negligible. Detailed behavior of crack propagation in one cycle is observed using the electrical potential drop technique. Figure 8.5(b) shows the change in the electrical potential in the cycles obtained, where the amounts of crack propagation during the rapid loading, Δa_U, and during the stress holding period, Δa_H, in one cycle can be separately evaluated.

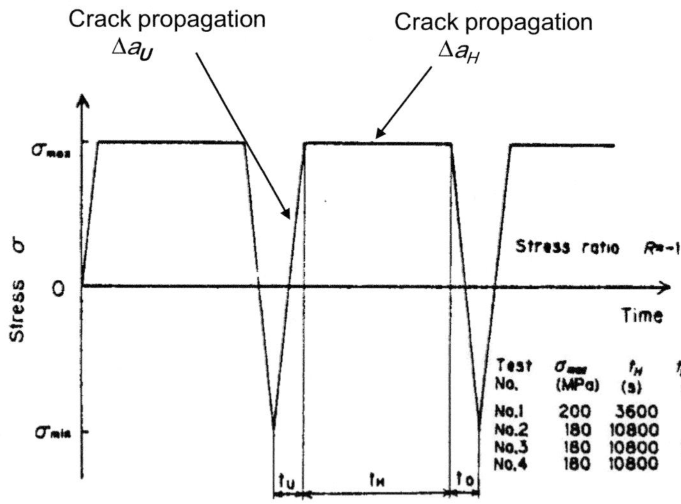

(a)

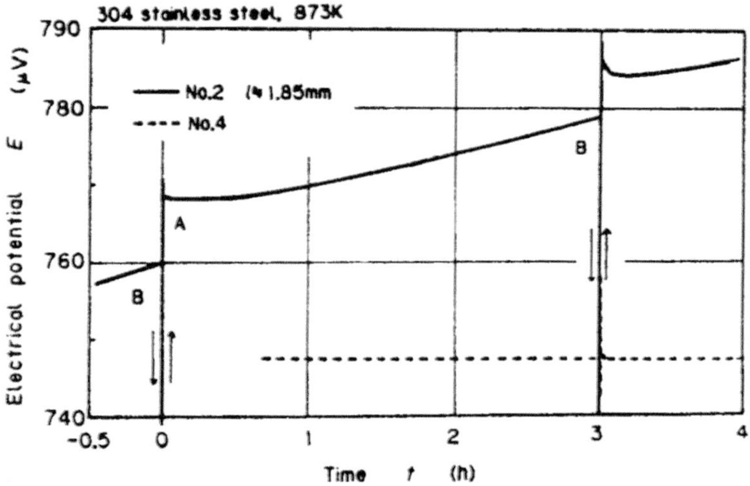
304 stainless steel, 873K
Electrical potential E (μV)
Time t (h)
No.2
No.4
A
B

(b)

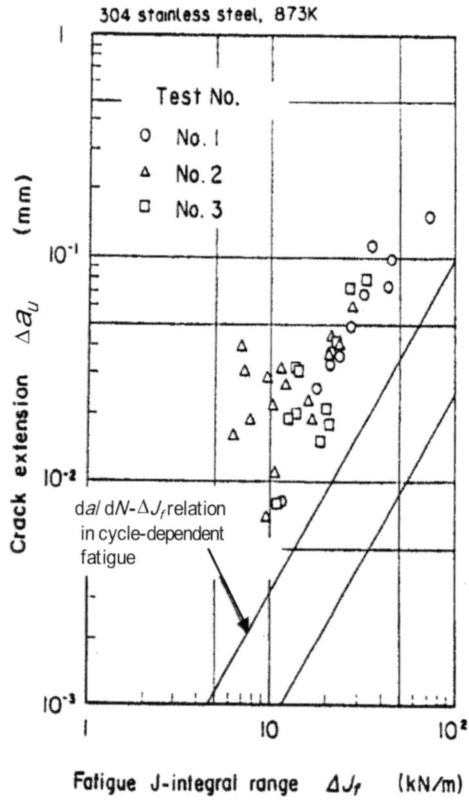
304 stainless steel, 873K
Test No.
No. 1
No. 2
No. 3
Crack extension Δa_u (mm)
da/ dN-ΔJ_f relation in cycle-dependent fatigue
Fatigue J-integral range ΔJ_f (kN/m)

(c)

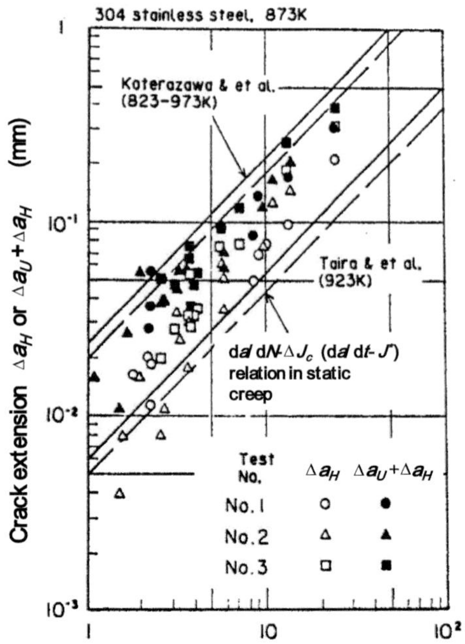

(d)

Figure 8.5 Continuous observation of crack propagation during one cycle by the electrical potential drop technique in time-dependent (creep-dominated) fatigue of 304 stainless steel at 873 K [Kitamura and Ohtani, 1986]. (a) Stress waveform consisted of tensile hold stress and spike-like compressive loading. (b) Change in the electric potential of No. 2 during cycles by crack propagation. No. 4 shows the result of a specimen with a dull notch indicating no crack propagation in the cycles. (c) Relationship between the crack extension during spike-like stress change in one cycle, Δa_U, and the fatigue J-integral range, ΔJ_f. (d) Relationship between the crack extension and the creep J-integral range. Solid marks: Relationship between the crack extension during one cycle, $\Delta a_U + \Delta a_H$, and the creep J-integral range, ΔJ_c. Follow marks: Relationship between the crack extension during stress hold period in one cycle, Δa_H, and the creep J-integral range, ΔJ_c.

Figure 8.5(c) shows the relationship between Δa_U and ΔJ_f, where the band represents the $da/dN–\Delta J_f$ relationship in cycle-dependent fatigue without creep influence, Eq. (7.1). The amount of crack propagation in the spike-like fatigue period is clearly larger than that in cycle-dependent crack propagation, suggesting significant acceleration of crack propagation due to creep damage formed near the crack tip during stress hold in the previous cycle.

Figure 8.5(d) shows the relationship of Δa_H and $\Delta a_U + \Delta a_H$ to ΔJ_c, where the band indicates the da/dN–ΔJ_c relationship in time-dependent (creep-dominated) fatigue, Eq. (7.2). The total amount of crack propagation during one cycle, $\Delta a_H + \Delta a_U$, aligns with the fracture mechanics law, supporting the independent crack propagation rule (Section 7.3.2(A)) in reality.

$$\frac{da}{dN} = \Delta a_U + \Delta a_H = C_c \Delta J_c \tag{8.5}$$

As seen from the above example, observation of crack propagation during one cycle provides fundamental information about the mechanism/mechanics of creep–fatigue interaction. Due to advances in observation equipment and technique, future research with even more precise experiments is expected.

8.3.2 Combination of Cycle- and Time-Dependent Crack Propagation

The result in the previous section indicates that the driving forces for crack propagation, ΔJ_c and ΔJ_f, and the amount of propagation are not synchronized in creep–fatigue interaction conditions. Expanding upon the behavior, various complex phenomena can be expected that have not been explored in detail.

Consider the simplest case that the loading condition is suddenly changed from time-dependent (creep-dominated) fatigue to cycle-dependent fatigue, as schematically depicted in Fig. 8.6(a). The crack passes through the creep-damaged zone ahead of the crack tip immediately after the change, resulting in a transient cracking behavior. Investigation of this behavior will provide information on the extent of damage process zone in front of the crack tip formed by the previous creep-dominated fatigue cycles.

This also signifies that crack propagation in creep and fatigue is non-linearly correlated, resulting in complex behavior under conditions in which cycle-dependent and time-dependent fatigue are sequentially combined as schematically illustrated in Fig. 8.6(b). Experiments based on systematic consideration will provide important clues to elucidate the complex mechanism of creep–fatigue interactions.

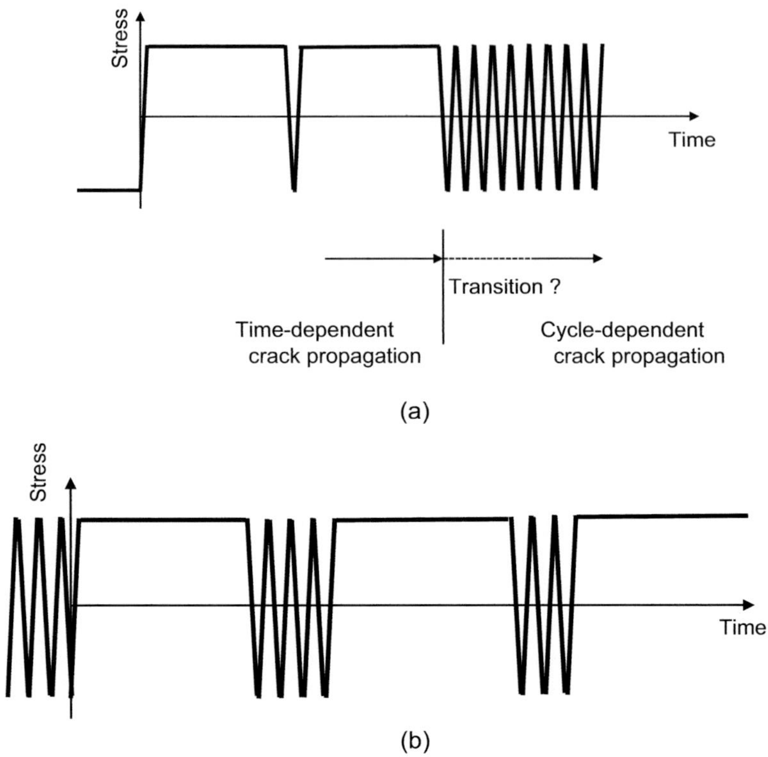

Figure 8.6 Combination of cycle-dependent fatigue and time-dependent fatigue. (a) Sudden change from time-dependent (creep-dominated) fatigue to cycle-dependent fatigue. (b) A sequential combination of cycle-dependent fatigue and time-dependent fatigue.

8.3.3 Simultaneous Crack Propagation due to Creep and Fatigue

In a general loading creep fatigue, both elastoplastic (time-independent) strain and creep (time-dependent) strain take place simultaneously, indicating that both ΔJ_c and ΔJ_f are present in the loading process. Although an evaluation method for ΔJ_c and ΔJ_f in such cases of a mixture is introduced in Section 7.5.3 and Appendix 7-A (Chapter 7), the singular fields have not been analytically investigated in detail. Strictly speaking, the physical nature of J-integral ranges in such cases is unknown.

Furthermore, although it becomes possible to estimate local "deformation" through ingenuity (for example, rapid straining method), there are still many unknowns regarding local "fracture" at the crack tip. This is one of the most important future challenges for understanding high temperature fracture mechanics.

8.3.4 Crack Propagation under Single Mechanism

Numerous experimental results demonstrate a clear distinction between crack propagation behavior in time-dependent fatigue and cycle-dependent fatigue, as schematically illustrated in Fig. 8.7(a).

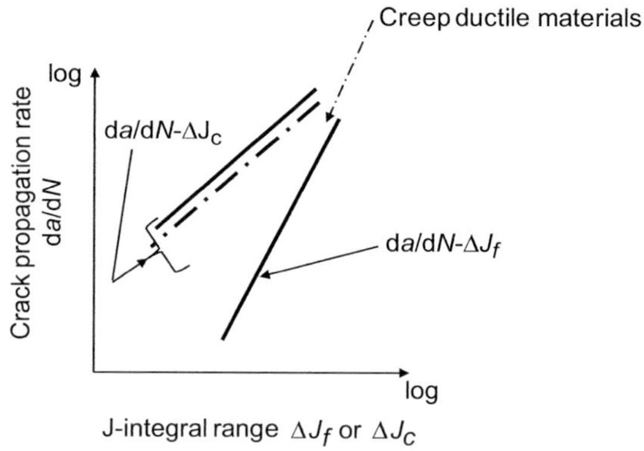

(a)

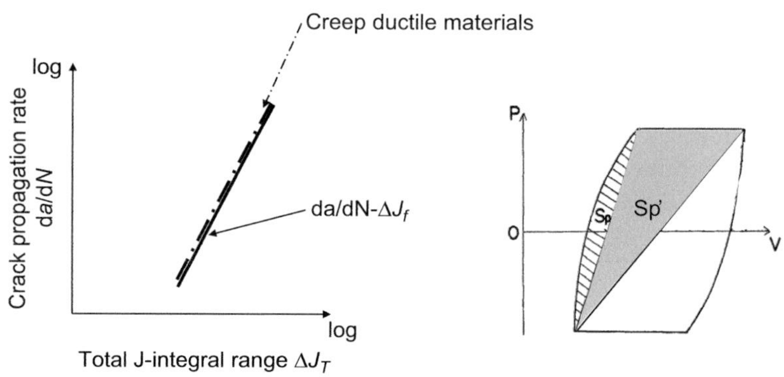

(b)

Figure 8.7 Crack propagation laws in cycle-dependent fatigue and time-dependent fatigue. (a) Case of standard steels and alloys. (b) Case of single fracture mechanism including creep.

However, for materials with high tensile ductility, this difference tends to become smaller. One fundamental question arises from it—Are there cases where the fracture mechanisms at the crack tip for both become the same depending on the conditions? If so, creep strain can be considered as time-dependent plastic strain, and there would be no need to divide the governing parameters of crack propagation into ΔJ_c and ΔJ_f. Similar to room temperature fatigue, da/dN should exhibit a good correlation with the fatigue J-integral range ΔJ_f, where creep strain is included in plastic deformation as illustrated in Fig. 8.7(b). Investigating this aspect will provide clues to elucidate the difference in the fracture mechanism at the crack tip.

References

Kawasaki, H., F. Ueno, K. Aoto, M. Ichimiya, and Y. Wada, 1992, Evaluation of long term creep-fatigue life for Type 304 stainless steel (Japanese), *Journal of The Society of Materials Science Japan*, 41, pp. 1773–1778. www.jstage.jst.go.jp/article/jsms1963/41/471/41_471_1773/_pdf/-char/ja

Kitamura T., H. Sugihara, and R. Ohtani, 1986, Effect of cyclic stress changes on creep crack propagation, *Journal of The Society of Materials Science Japan*, 35, pp. 260–266. https://www.jstage.jst.go.jp/article/jsms1963/35/390/35_390_260/_pdf/-char/ja

Kitamura T., and R. Ohtani, 1986, Crack propagation under interaction between creep and fatigue, (in Japanese), *Transaction of Japanese Society of Mechanical Engineers A*, 52, pp. 1816–1823. https://www.jstage.jst.go.jp/article/kikaia1979/52/480/52_480_1816/_pdf/-char/ja

Ohtani R., T. Kitamura, and W. Zhou, 1995, Effect of internal creep microcracks on fatigue macrocrack propagation at high temperatures, *Materials Science Research International*, 1, pp. 238–246.

Chapter 9

Propagation of Microstructurally Small Cracks

9.1 Influence of Microstructure on Crack Propagation

9.1.1 Effect of Large Microstructure on Macro Crack Propagation

Figure 9.1(a) shows the propagation behavior of macro crack in time-dependent (creep-dominated: cp-type) fatigue of Inconel MA754 (mechanical alloy) with a microstructure of large elongated grains at 1273 K [Ohtani *et al.*, 1991]. The average size of grains is a few mm in the loading direction and 300 mm in the cross section, respectively. The crack propagates through grains in a step-like manner, repeating propagation and stagnation. The broken lines in the figure indicate grain boundaries perpendicular to the crack, elucidating that they significantly inhibit crack propagation. The curve also shows that the influence is strong for a shorter crack and disappears for one longer than 2.5 mm which is about 8 times the representative scale of the microstructure. As a result, the relationship between da/dN

High Temperature Fracture Mechanics
Takayuki Kitamura
Copyright © 2025 Jenny Stanford Publishing Pte. Ltd.
ISBN 978-981-5129-55-7 (Hardcover), 978-1-003-65328-8 (eBook)
www.jennystanford.com

and ΔJ_c has large fluctuations as shown in Fig. 9.1(b), signifying that the microstructure has strong influence on the fracture mechanics laws for short crack.

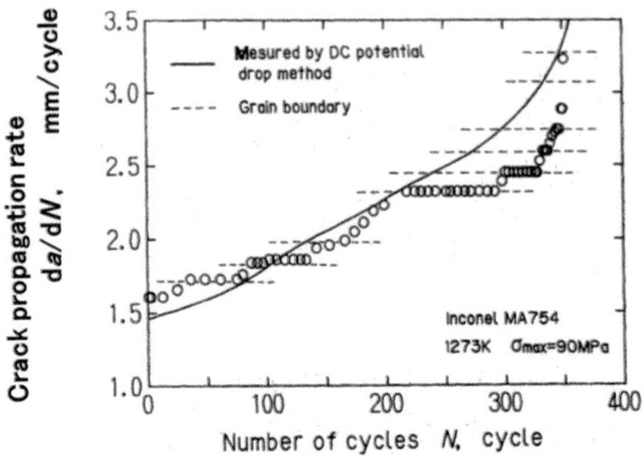

(a)

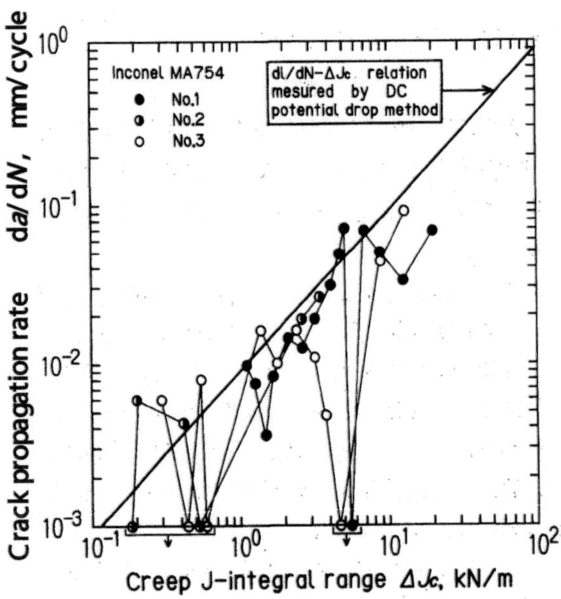

(b)

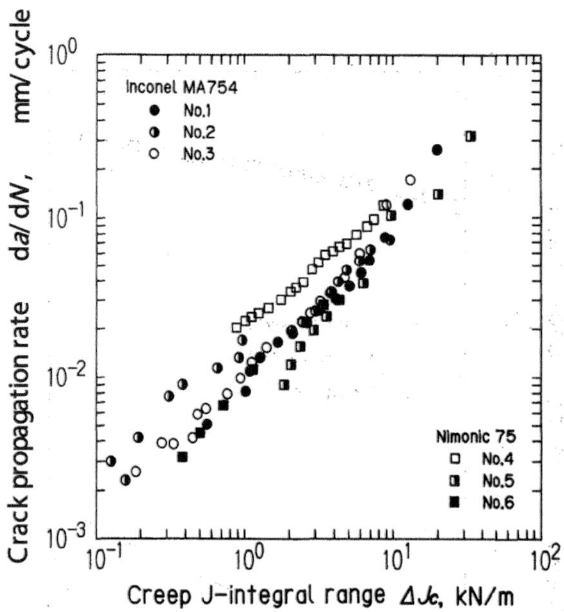

(c)

Figure 9.1 Crack propagation behavior of Inconel MA754 consisted of coarse elongated grains in creep-dominated (time-dependent) fatigue at 1273 K [Ohtani *et al.*, 1991]. (a) Crack propagation curve where the crack length is measured by means of a microscope, and the dashed lines indicate the grain boundaries. The solid line shows the curve obtained by the electrical potential drop technique. (b) Relationship between the crack propagation rate on specimen surface, da/dN, and the creep J-integral range, ΔJ_c. (c) Relationship between the averaged crack propagation rate along the crack front, da/dN, and the creep J-integral range, ΔJ_c.

The above result is based on the crack length measured at the surface by means of a microscope, while it is a through crack in a plate specimen with the thickness of 5 mm. The solid line in Fig. 9.1(a) presents the crack length measured by the electrical potential drop technique, indicating the average crack length along the crack fronts (50 mm × 2) of the specimen thickness. The fluctuations in the relationship clearly disappear, meaning continuous crack propagation by averaging the microstructural influence. Figure 9.1(c) shows the relationship between da/dN and ΔJ_c based on the crack length measured by the electrical potential drop technique, indicating that the fracture mechanics law holds its effectiveness. In other words, the fracture mechanics laws are based on homogeneous approximation where the crack (or crack front) is long enough in comparison with the size of microstructure.

9.1.2 Propagation of Microstructurally Small Cracks

In static creep of smooth specimens including creep–fatigue conditions, a huge number of small cracks appear on the entire surface with high density (for example, Fig. 9.4(a)). Figure 9.2(a) depicts the propagation properties of small cracks initiated on the surface of a smooth specimen in static creep of 304 stainless steel with a grain diameter of 500 mm at 973 K [Ohtani *et al.*, 1984]. The figure shows propagation of four typical cracks extracted from numerous small cracks evaluated from *in situ* observation, and the grain boundary features near them are illustrated on the right side. The cracks exhibit prominent fluctuations and are strongly influenced by the microstructure. While they propagate smoothly along grain boundaries, the propagation is suppressed near the triple-junctions/bends of grain boundaries. Figure 9.2(b) shows the crack propagation rate, where the band in the figure indicates the $\mathrm{d}a/\mathrm{d}t$–a relationship of macro crack converted from the $\mathrm{d}a/\mathrm{d}t$–J^* relationship, assuming a crack in an infinite homogeneous creep body ($J^* \propto a$, Eq. (3.14)). Microstructural effects bring about prominent fluctuations where the propagation of small cracks far exceeds the range of the band.

Figure 9.3 shows the $\mathrm{d}a/\mathrm{d}t$–J^* relationship of many small cracks initiated on the surface of a smooth specimen in static creep of 316 stainless steel at 923 K [Ohtani *et al.*, 1987]. The $\mathrm{d}a/\mathrm{d}t$ exhibits fluctuations of more than 100 times for the same crack length, and the average fluctuations of small cracks almost matches the fracture mechanics law (the $\mathrm{d}a/\mathrm{d}t$–J^* relationship) of macro crack. In other words, the characteristics of macro crack present the averaging of the local fluctuation of propagation at crack front due to microstructure.

It should be noted that the cracks initiated are not only "short" in the length on the surface but also "small," indicating short in the depth as well. Thus, they are termed "small cracks" distinguished from "short cracks."

The crack path brings about microstructural influence, meaning that each grain boundary possesses its own strength (fracture resistance). On the other hand, each crystal grain exhibits

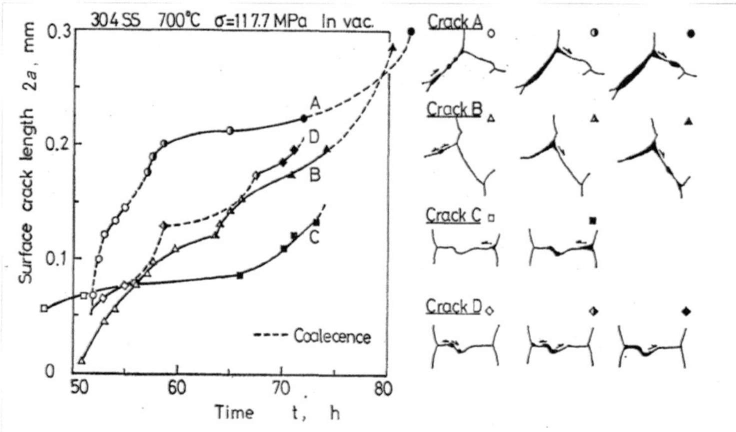

(a)

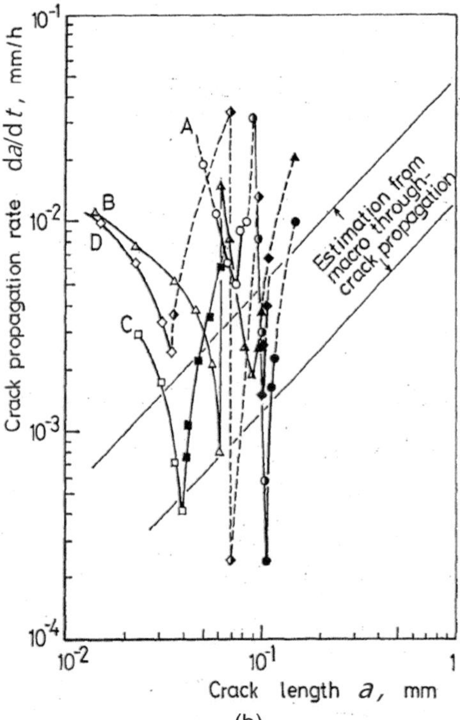

(b)

Figure 9.2 Propagation behavior of small cracks initiated on surface of smooth specimen in static creep of 304 stainless steel at 973 K [Ohtani *et al.*, 1984]. (a) Propagation curves of four typical small cracks and illustration of their microstructural morphology with grain boundaries. (b) The crack propagation rate of microstructurally small cracks.

different deformation behavior under uniaxial creep loading due to differences in crystallographic orientation and network (shapes) of poly-grains. This variation leads to fluctuation in the magnitude of J^* (fracture driving force). Therefore, it should be noted that the influence of microstructure consists of two factors—deformation factor (on J^*) and fracture factor (on C_c). Both of them are included in the fluctuation of the da/dt–J^* relationship shown in Fig. 9.3.

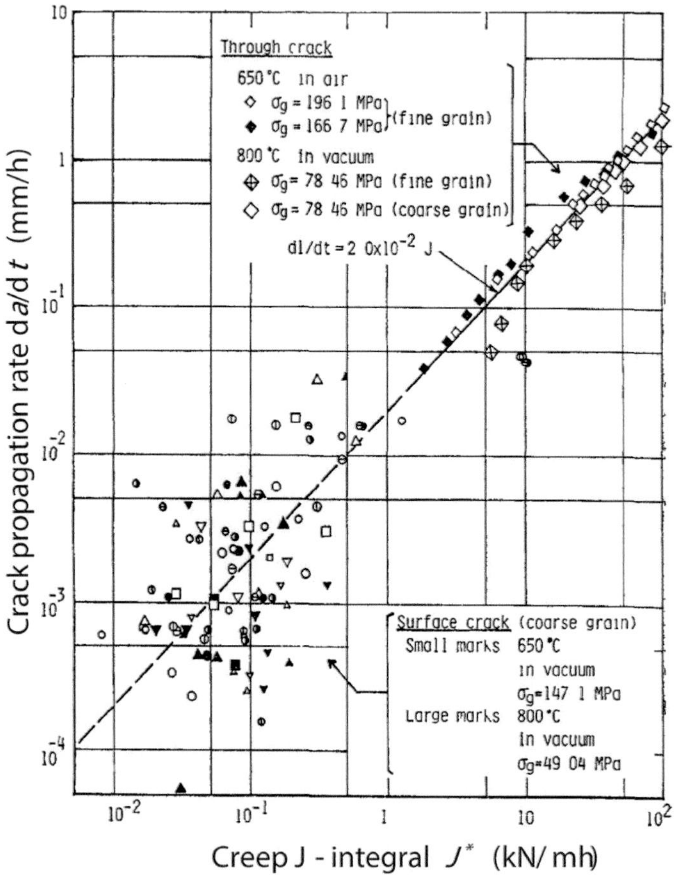

Figure 9.3 Relationships between the crack propagation rate, da/dt, and the creep J-integral, J^*, of microstructurally small cracks in static creep of 316 stainless steel at 923 K in comparison with those of macro crack (solid line) [Ohtani *et al.*, 1982].

9.2 Propagation of Small Cracks in High Temperature Fatigue

9.2.1 Initiation and Propagation of Microstructurally Small Cracks in Time-Dependent (Creep-Dominated) Fatigue

In high temperature fatigue, small cracks initiate in the very early stages, often less than 10% of failure life, and their growth leads to the failure of the entire component [Sakurai *et al.*, 1986]. Figure 9.4(a) illustrates the typical morphology of small cracks distributed on the surface of smooth specimen observed in time-dependent (creep-dominated) fatigue of 304 stainless steel at 923 K in a vacuum [Tada *et al.*, 1990]. It should be noted that no cracks are observed inside the specimen even after failure.

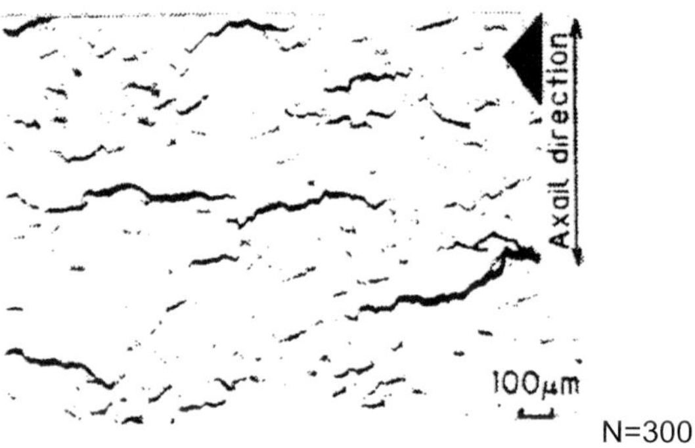

(a)

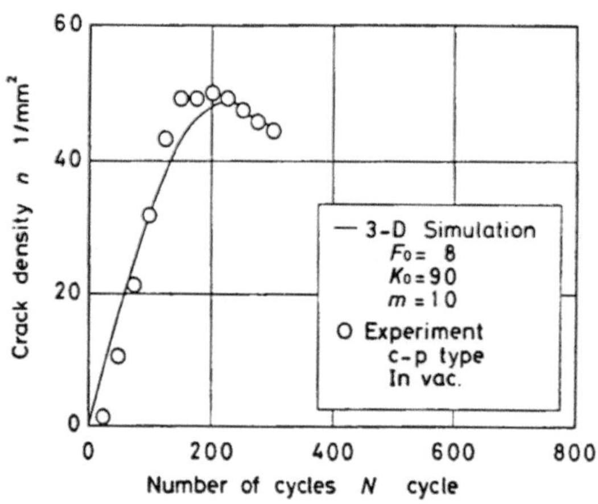

(b)

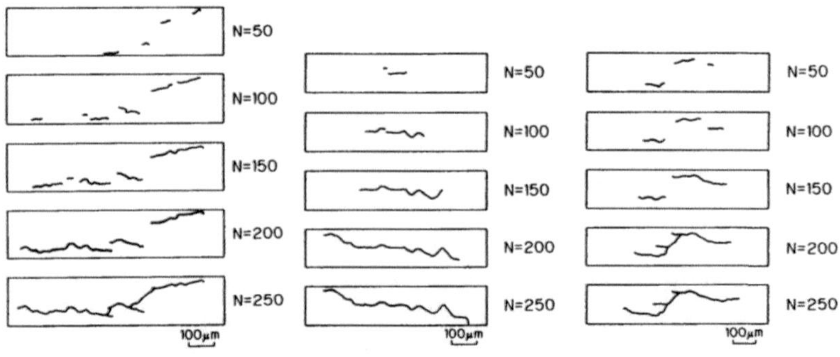

(c)

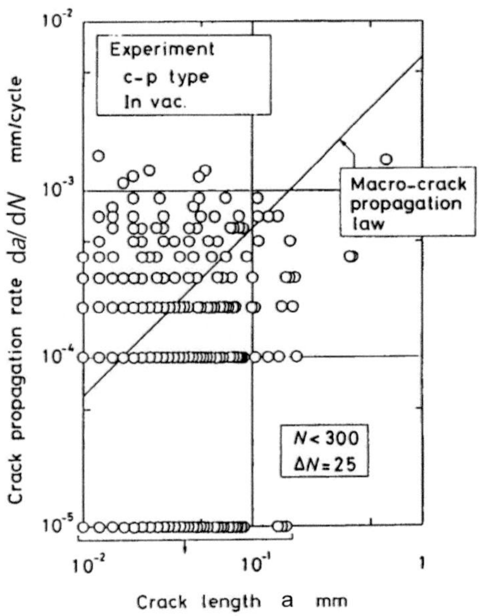

(d)

Figure 9.4 Propagation behavior of microstructurally small cracks initiated on the surface of smooth specimen in cp-type time-dependent (creep-dominated) fatigue of 304 stainless steel at 923 K in a vacuum. (a) Morphology of microstructurally small cracks on the specimen surface at $N = 300$ cycles [Ohtani *et al.*, 1986]. (b) The crack density defined as number of cracks per unit area [Tada *et al.*, 1990]. (c) Crack propagation behavior of several cracks [Kitamura *et al.*, 1990]. (d) Relationship between the crack propagation rate, da/dN, and the crack length, a [Tada *et al.*, 1990].

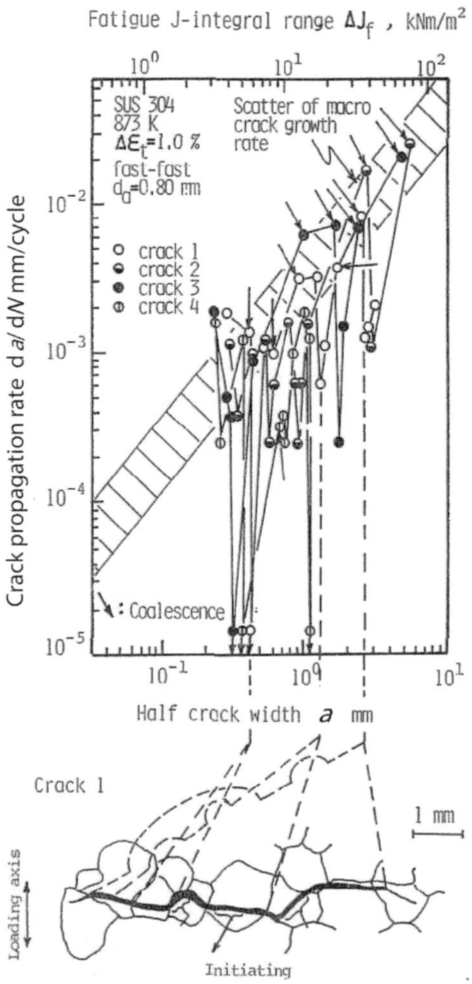

Figure 9.5 Propagation behavior of microstructurally small cracks initiated on surface of sooth specimen in fatigue of 304 stainless steel at 823 K [Okazaki *et al.*, 1987].

The areal density of surface small cracks increases with creep–fatigue cycles, as shown in Fig. 9.4(b). The initiation life differs for each small crack, evidently indicating strong influences of microstructure. The decrease in the density in the late stage of life is attributed to the frequent coalescence of small cracks due to very high density.

Focusing on specific cracks in Fig. 9.4(a), the crack propagation behavior and the da/dN–a relationship are depicted in Fig. 9.4(c) and (d), respectively [Ohtani *et al.*, 1986]. Similar to the case of static creep, da/dN shows large fluctuations for the same crack length, and there are clearly non-propagating cracks. The straight line in the figure represents the da/dN–ΔJ_c relationship under this creep–fatigue condition, converted using the J–a relationship. The fluctuation of microstructurally small cracks centers on the characteristic of macro crack, meaning that the fracture mechanics law is based on the average of microstructural influence.

9.2.2 Propagation of Microstructurally Small Cracks in Cycle-Dependent Fatigue

Figure 9.5 shows the propagation behavior of small cracks initiated on the surface of a smooth specimen in cycle-dependent fatigue of 304 stainless steel with grain diameter of 0.8 mm at 873 K [Okazaki *et al.*, 1987 and 1989]. The fluctuation of da/dN is prominent, and there are non-propagation cracks, exhibiting similar characteristics to those observed in time-dependent (creep-dominated) fatigue. Their average da/dN corresponds well with the da/dN–ΔJ_f relationship of macro crack indicated by the band in the figure.

9.2.3 Influences of Air Environment

Experiments are conducted in a vacuum (about 10^{-2} Pa) and in air under time-dependent (creep-dominated) fatigue of 304 stainless steel at 923 K [Ohtani *et al.*, 1986]. Oxide adhesion is observed around the small cracks in the air environment, displaying a clear different morphology from that in the vacuum. However, as shown in Fig. 9.6, the fluctuation behavior in the da/dN–a relationships in air is similar to that in a vacuum, and their average da/dN is governed by ΔJ_C as well. Surprisingly, the air environment presents a weak influence on the propagation of microstructurally small cracks in terms of fracture mechanics laws. Here, it should be noted that oxidation has a strong influence on crack density (crack initiation).

As the effect depends on materials and experimental conditions, it is an important issue for future research.

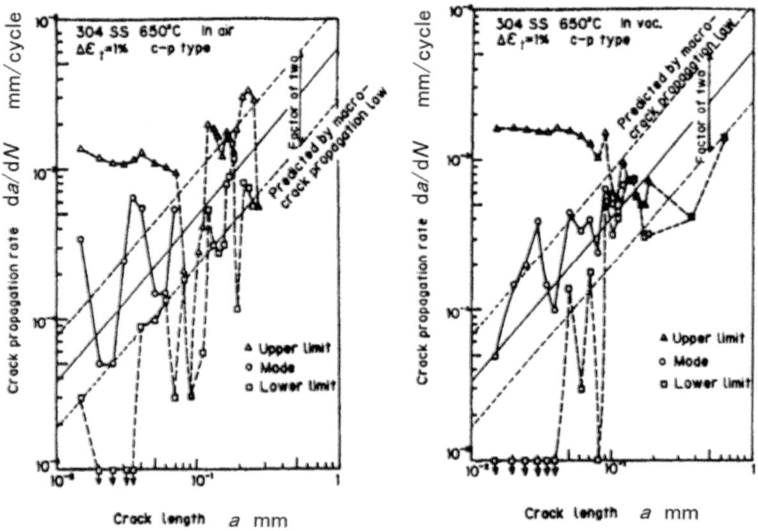

Figure 9.6 Propagation behavior of microstructurally small cracks in time-dependent (creep-dominated) fatigue of 304 stainless steel in air and in a vacuum environment at 923 K [Ohtani *et al.*, 1986].

9.2.4 Influences of Compressive Creep

Experiments are conducted in time-dependent fatigue with different compressive strain rates as depicted in Fig. 9.7(a) [Kitamura *et al.*, 1990]. Figure 9.7(b) shows the propagation behavior of microstructurally small cracks, elucidating weak influence. On the other hand, the number of cracks initiated (crack density) is strongly influenced by the compressive strain rate as shown in Fig. 9.7(c), resulting in different coalescence characteristics. This subtle mechanism of damage healing by compressive creep brings about the stress (strain) waveform effects on the life in high temperature fatigue.

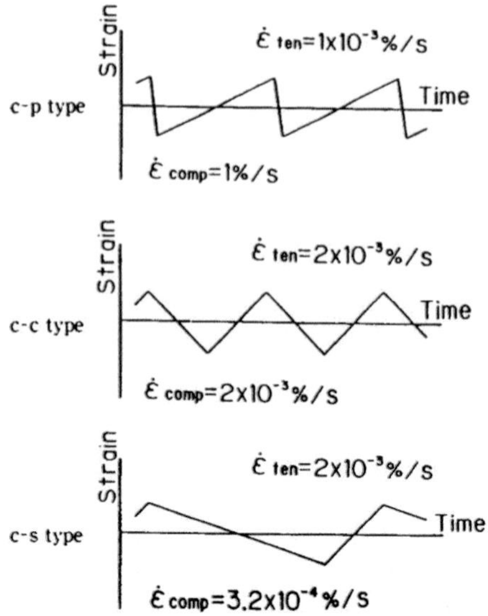
Strain
Ɛ̇ ten=1x10⁻³%/s
Time
c-p type
Ɛ̇ comp=1%/s
Strain
Ɛ̇ ten=2x10⁻³%/s
Time
c-c type
Ɛ̇ comp=2x10⁻³%/s
Strain
Ɛ̇ ten=2x10⁻³%/s
Time
c-s type
Ɛ̇ comp=3.2x10⁻⁴%/s

(a)

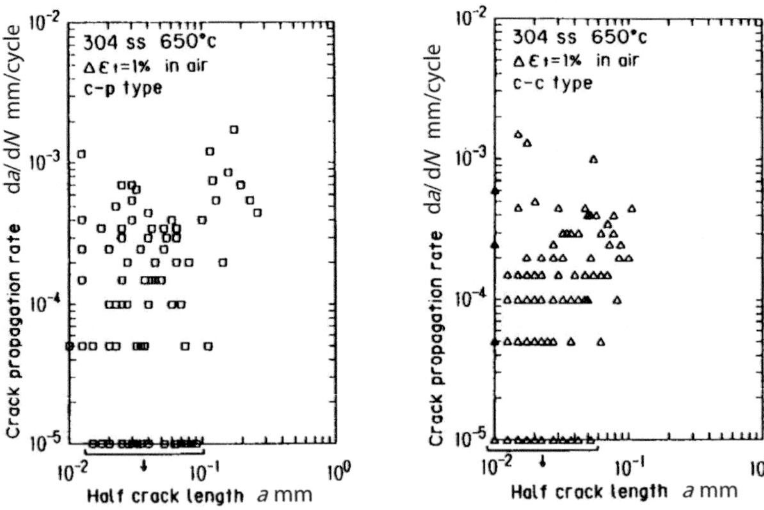
304 ss 650°c
ΔƐt=1% in air
c-p type
Crack propagation rate da/dN mm/cycle
Half crack length a mm
304 ss 650°c
ΔƐt=1% in air
c-c type
Crack propagation rate da/dN mm/cycle
Half crack length a mm

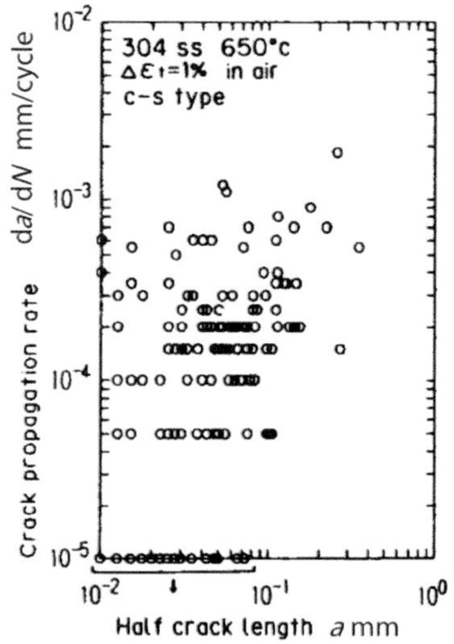

(b)

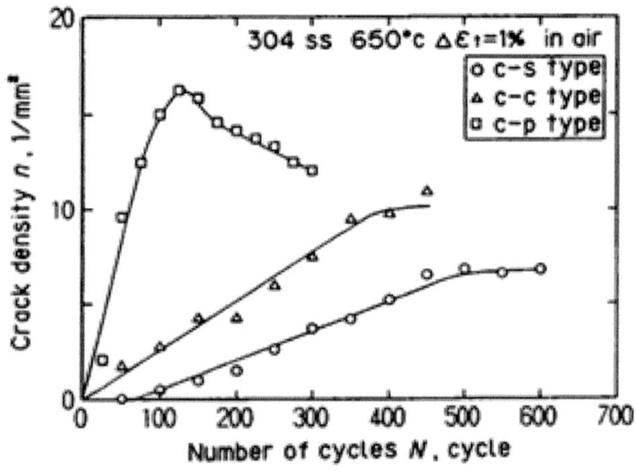

(c)

Figure 9.7 Influence of compressive creep on propagation of microstructurally small cracks in time-dependent (creep-dominated) fatigue of 304 stainless steel at 923 K [Kitamura *et al.*, 1990]. (a) Strain waveforms applied. (b) Influence of compressive creep on relationship between the crack propagation rate, d*a*/d*N*, and the crack length, *a*. (c) Influence of compressive creep on the crack density.

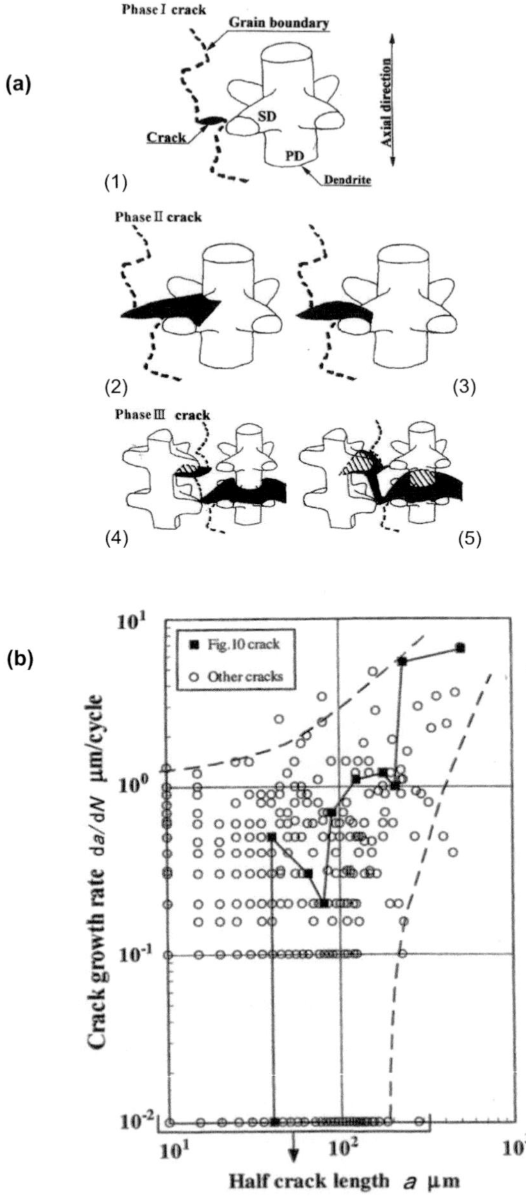

Figure 9.8 Initiation and propagation of microstructurally small crack in time-dependent (creep-dominated) fatigue of directionally solidified nickel base superalloy, Mar-M 247, at 1273 K [Okada *et al.*, 1998]. (a) Illustrations of crack propagation process of small crack from (1) to (5). (b) Relationship between the crack propagation rate, d*a*/d*N*, and the crack length, *a*.

9.2.5 Crack Propagation in Microstructure-Controlled Materials

Materials with various advanced microstructures have been developed for high temperature applications. Small cracks, of course, exhibit characteristic behavior due to the geographic and crystallographic features of such microstructures.

As an example, Fig. 9.8(a) illustrates the forming process of the main crack observed in a unidirectionally solidified coarse-grained nickel-base superalloy Mar-M247 in time-dependent (creep-dominated) fatigue [Okada *et al.*, 1998]. The crack initiates at a grain boundary bend on the surface of smooth specimen and propagates into a grain. Inside the grain, as a dendrite blocks its progress, the crack propagates around it. Finally, it cuts through the dendrite after circumventing it. The crack propagation rate shows similar fluctuation behavior as shown in Fig. 9.8(b).

Highly functional materials have been developed by controlling the microstructure. Although some experiments have been conducted on these, there is no unified mechanical understanding.

9.3 Transition from Microstructurally Small Crack to Macro Crack

Figure 9.9 shows the da/dN–ΔJ_c relationship in time-dependent fatigue with different applied strain ranges, where the straight line represents the relationship of a macro crack [Tada *et al.*, 1992]. In the figure, the symbols change depending on the crack length.

Test (1) Crack length is in the range of 30 mμ – 100 μm
Test (2) Crack length is in the range of 100 μm – 1 mm
Test (3) Crack length is longer than 1 mm

Test (3) is clearly classified into macro crack, since ΔJ_c governs da/dN without the fluctuation. The influence of microstructure is sufficiently averaged out. Test (1) exhibits a large fluctuation and many of them stop propagating due to the microstructural influence, which is categorized into microstructurally small crack.

The representative length of the microstructure is the average grain size, which is about 50 mm in this material. The minimum length for a macro crack is about 10 times the representative length of microstructure. In other words, if it is shorter than that, it will be strongly influenced by the microstructure.

(i) $a < 2d$ Microstructurally small crack
(ii) $2d < a < 10d$ Transition
(iii) $10d < a$ Macro crack

The dependence of applied strain (stress) range is schematically illustrated in Fig. 9.9(b).

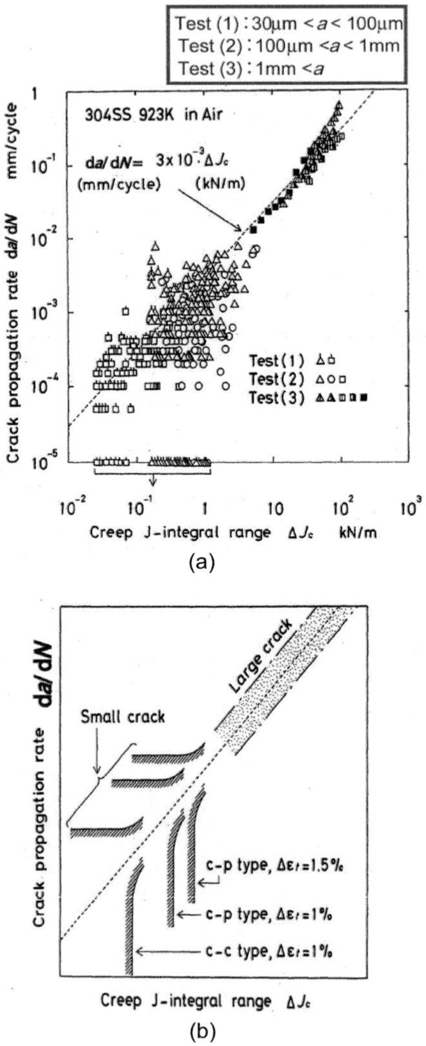

Figure 9.9 Transition from microstructurally small crack to macro crack in time-dependent (creep-dominate) fatigue of 304 stainless steel at 923 K in air [Tada *et al.*, 1992]. (a) Experimental results. (b) Schematic diagram of transition from microstructurally small crack to macro crack.

9.4 Statistical Feature of Initiation and Propagation: Modelling and Monte Carlo Simulation

In creep-dominated conditions, the orthodox way to explore fluctuation due to microstructure is to identify the differences in deformation behavior of each grain related to the creep J-integral (fracture driving force), and the differences in the creep strength of each grain boundary related to C_c in the da/dN–ΔJ_c relationship (fracture resistance). However, due to experimental and analytical difficulties, the challenge remains for future works. This book introduces a semi-mechanical method that incorporates the influence of microstructure and provides an overall scope of complexity due to random behavior of numerous small cracks.

9.4.1 Modelling of Propagation of Microstructurally Small Cracks

The Monte Carlo method, which handles the random nature of microstructure on small cracks in creep, is proposed relatively early in the investigation, in order to simulate complex behavior [Ohtani and Kitamura, 1987].

The following is simulation method, which addresses crack propagation along grain boundaries in time-dependent fatigue [Ohtani *et al.*, 1988]. Furthermore, applying this method can be readily extended to transgranular crack propagation in cycle-dependent fatigue.

(A) Preparation of material microstructure

Consider a grain boundary network on the surface of polycrystalline material as illustrated in Fig. 9.10(a). It is possible to reproduce a realistic network with geometric randomness of microstructure of a target material through numerical simulation [Ohtani *et al.*, 1988]. A fracture resistance value, R, is set for each grain boundary, where R is a random variable varying for each grain boundary. The simulation in the following section adopts the simplest uniform probability distribution ranging from 0 to 1 for R.

(B) Crack initiation: Fig. 9.10(b)

When there is no crack at adjacent grain boundaries (Fig. 9.10(b)), the fracture resistance of the grain boundary, R, decreases by F (driving force for crack initiation) in each creep–fatigue stress (strain) cycle. This process is repeated until $R - \Sigma F$ reaches zero, which is defined as the crack initiation at the grain boundary. Here, F is a function of the normal stress on the grain boundary, assuming to be simply determined by the angle between the stress axis and the grain boundary. Since the local normal stress at the grain boundary differs depending on the deformation property of surrounding grains and constraint imposed by them, F is a random variable. However, the probability can be convolved into a single variable in the subtraction calculation process (R–NF), so that the probability of crack initiation is determined solely by the fluctuation of either R or F [Kitamura and Ohtani, 1990]. F is treated as a deterministic variable dependent on the applied stress and the grain boundary angle against the stress axis, assuming R involves both randomness. Furthermore, F can be standardized using R as a reference scale, ranging from 0 to 1.

(C) Crack propagation: Fig. 9.10(c)

When a crack exists at the adjacent grain boundary (Fig. 9.10(c)), R decreases by $K_o a$ (driving force for crack propagation) in each creep–fatigue cycle, $R - K_o a$, considering the proportionality of J^* to the crack length for small cracks, Eq. (3.14). This process is repeated until the residual resistance reaches zero, which is defined as the crack extension of one grain boundary. Similar to process (B), K_o is a deterministic variable obtained by the da/dN–ΔJ_c relationship of macro crack (average of microstructural effect on crack propagation).

This model has an extremely simple structure with only a random variable R, which follows a probability distribution function ranging from 0 to 1, and two deterministic variables F and K_o. Here, it should be noted that the fitting parameter is only F for the crack initiation.

This simplified model can surprisingly reproduce the complex behavior of microstructurally small cracks as shown in the next section.

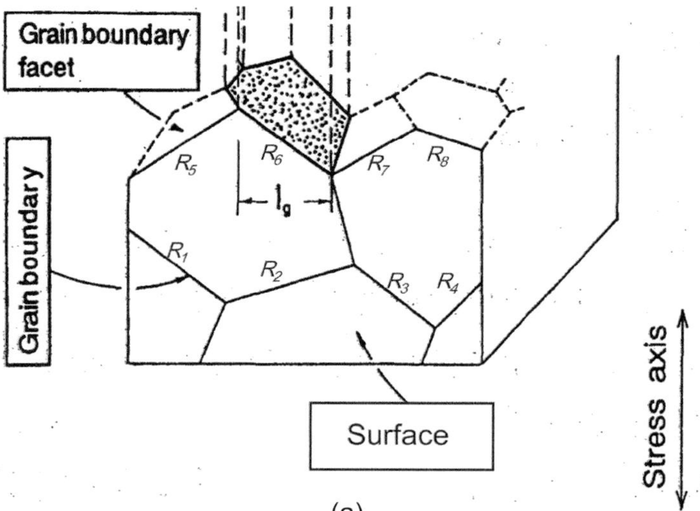
Grain boundary facet
Grain boundary
R5
R6
R7
R8
R1
R2
R3
R4
lg
Surface
Stress axis

(a)

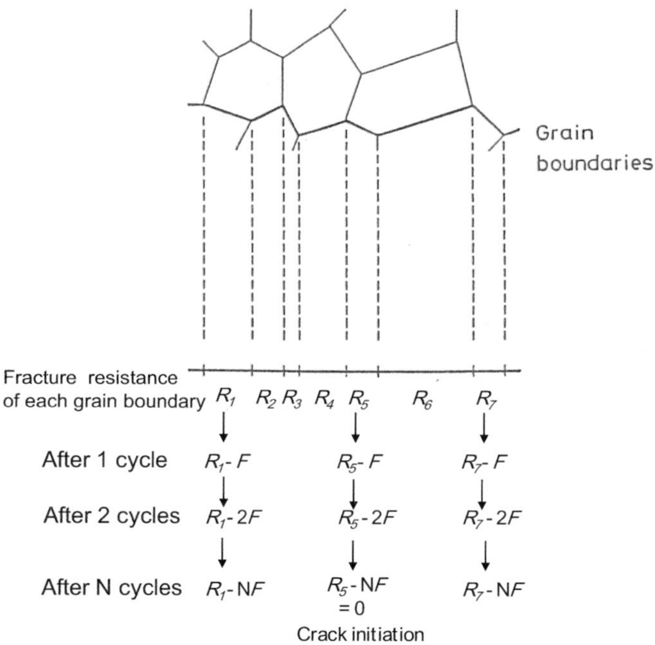
Grain boundaries
Fracture resistance of each grain boundary
R1 R2 R3 R4 R5 R6 R7
After 1 cycle
R1- F
R5- F
R7- F
After 2 cycles
R1- 2F
R5- 2F
R7- 2F
After N cycles
R1- NF
R5- NF = 0
Crack initiation
R7- NF

(b)

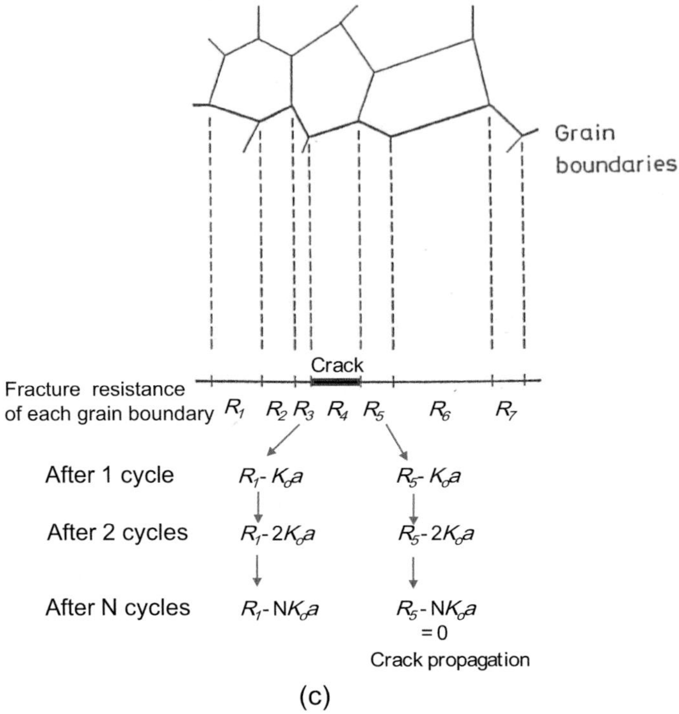

Figure 9.10 Illustrations explaining numerical simulation model of initiation and propagation of microstructurally small cracks. (a) Network of grain boundaries on surface. (b) Resistance and driving force in crack initiation process. (c) Resistance and driving force in crack propagation process.

9.4.2 Comparison of Cracking Feature with Experimental Results

Figure 9.11 presents the results of simulation and experiment (see Fig. 9.4) under the same condition for comparison [Tada *et al.*, 1990], depicting the crack propagation rate, the crack density, the distribution of crack length, and morphology of small cracks on the specimen surface, respectively. In Figs. 9.4(a) and (b), the marks represent the experimental results, while the solid lines indicate simulation results. The simulation statistically reproduces the random characteristics of initiation and propagation of microstructurally small cracks, including the transition to macrocrack.

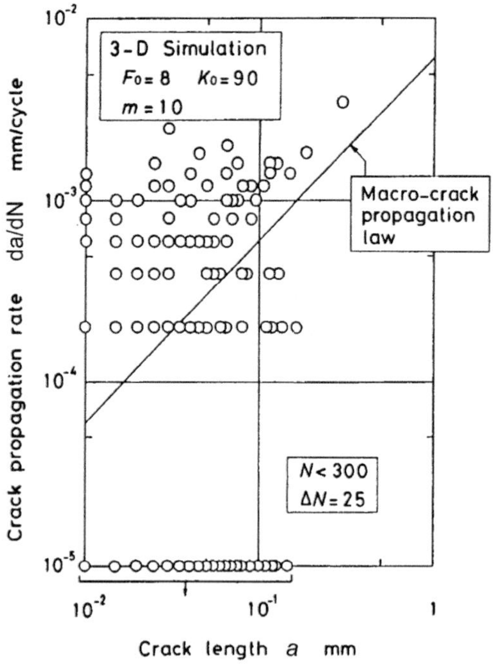
3-D Simulation
F0 = 8 K0 = 90
m = 10
Macro-crack propagation law
N < 300
ΔN = 25
Crack propagation rate da/dN mm/cycle
Crack length a mm
10^-2
10^-3
10^-4
10^-5
10^-1
1

(a)

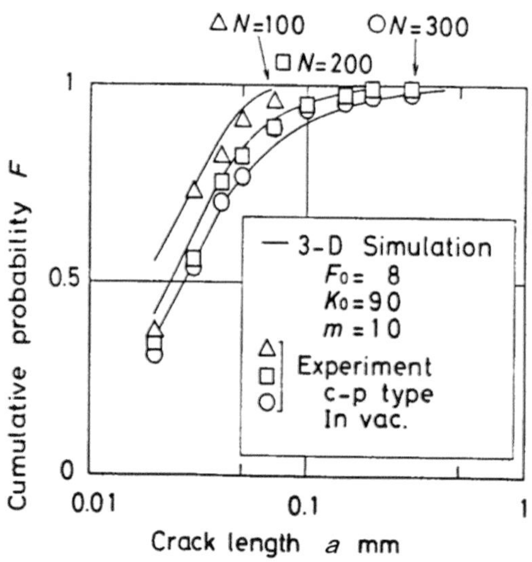
ΔN=100
□N=200
○N=300
— 3-D Simulation
F0 = 8
K0 = 90
m = 10
Experiment
c-p type
In vac.
Cumulative probability F
Crack length a mm
1
0.5
0
0.01
0.1
1

(b)

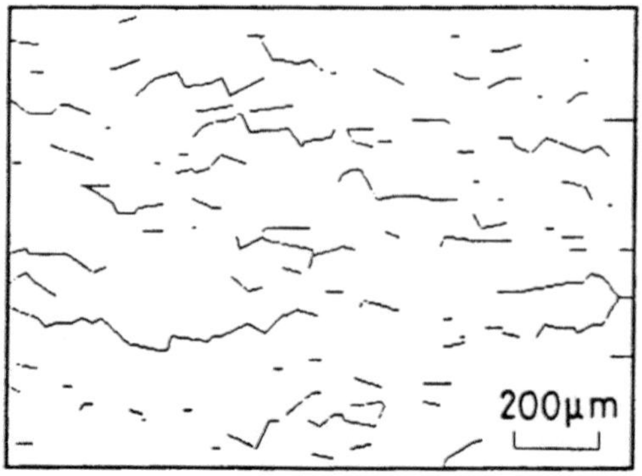

(c)

Figure 9.11 Comparison of behavior of small crack initiation and propagation in cp-type time-dependent (creep-dominated) fatigue of 304 stainless steel at 923 K obtained by simulation and experiment [Tada *et al.*, 1990]. (a) Relationship the crack propagation rate, d*a*/d*N*, and the crack length, *a*. (b) Change in cumulative probability of crack length. (c) Morphology of distributed small cracks on surface.

The success of simple model suggests the existence of a simple basic principle behind the complex phenomenon of initiation and propagation of each crack.

The main points drawn from these results are summarized below:

- Randomness due to microstructure is approximately expressed by a uniform probability distribution.
- The driving forces (F and K_0a) for crack initiation and crack propagation are significantly different.
- Crack initiation is dominated by normal stress on the grain boundary.
- Crack propagation follows the fracture mechanics laws on average.

9.5 Small Cracks Initiated within Material

In the case of normal time-dependent (creep-dominated) fatigue in smooth materials, small cracks typically initiate on the surface. However, under conditions characterized by solely tensile creep (cp-type fatigue or static creep) with very-low stress, a significant number of small cracks initiate within the material [Zhou *et al.*, 1995]. Figure 9.12 shows an example of small cracks initiated within

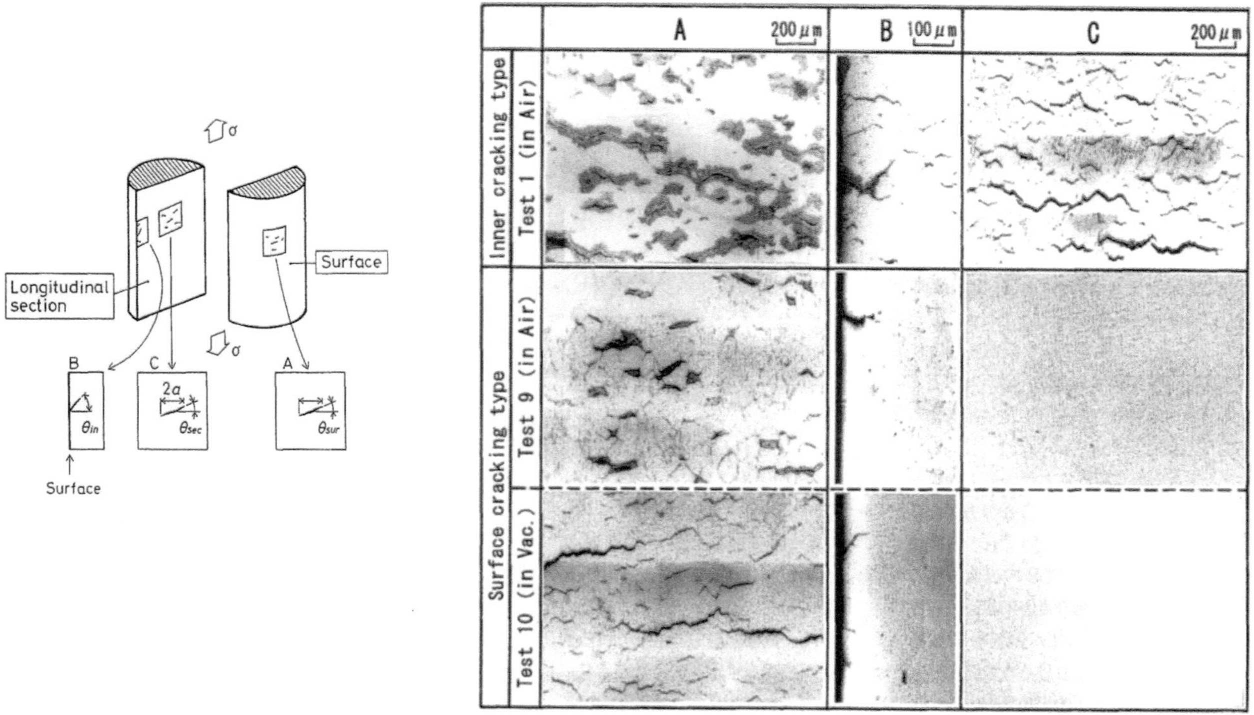

Figure 9.12 Morphology of small cracks on cross section of specimen under cp-type time-dependent (creep-dominated) fatigue of 304 stainless steel [Zhou *et al.*, 1995]. Inner small cracks appear on the cross section of specimen in creep fatigue at 1073 K. Only surface cracks appear in creep fatigue at 923 K.

a 304 stainless steel subjected to cp-type time-dependent fatigue at 1073 K. It should be noted that small cracks initiate only on the surface in time-dependent fatigue with tension and compression creep (cc-type), even at such very-high temperatures (low stresses). This suggests that damage accumulated within the material due to tensile creep is healed by compressive creep.

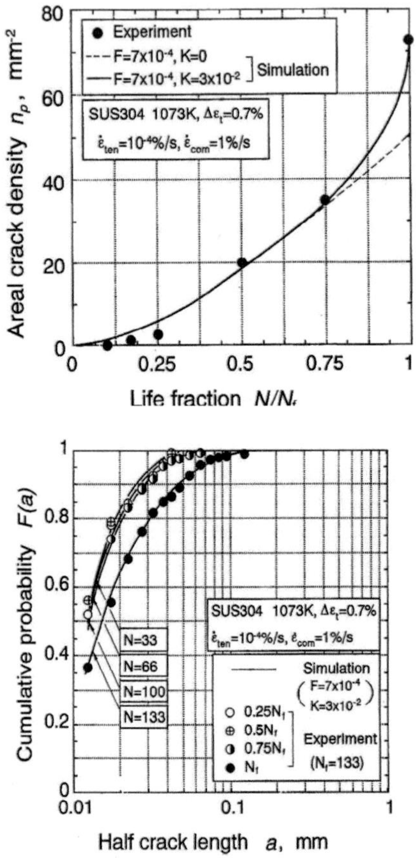

Figure 9.13 Change in cumulative probability of the crack length on the cross section of specimen under cp-type time-dependent (creep-dominated) fatigue of 304 stainless steel at 1073 K [Tada *et al.*, 1995].

It is possible to measure the length distribution from cross sections of multiple stopped specimens subjected to different numbers of time-dependent fatigue cycles [Tada *et al.*, 1995].

Figure 9.13 shows how the internal crack length distribution changes with the number of fatigue cycles. Although crack sizes increase as fatigue progresses, extracting direct crack propagation characteristics from this data is challenging. Performing a three-dimensional simulation inside the specimen to replicate the crack length distribution and crack density using the method described in the previous section, F and K_0 can be determined by inverse analysis. The simulation results indicated by solid lines imply that the summary in the previous section remains effective for internal cracking, including the fracture mechanics laws.

Experimental verification is currently insufficient, but with modern advances in observing methods of internal cracks, it is anticipated that non-destructive continuous observation will become feasible. These are issues that remain for the future. Furthermore, the scenario of numerous small cracks initiating and propagating throughout the volume aligns well with the concept of damage mechanics. The success of simple simulations supports its effectiveness, though detailed observation of the fracture process is crucial before conducting the analysis.

References

Kitamura T., N. Tada, M. Abe, M. Yumita, and R. Ohtani, 1990, Effect of compression: Going strain rate on initiation and growth of small cracks under creep-fatigue condition, *Transaction of JSME, Ser. A*, 56, pp. 575–581. https://www.jstage.jst.go.jp/article/kikaia1979/56/523/56_523_575/_pdf/-char/ja

Ohtani R., M. Okuno and R. Shimizu, 1982, Microcrack growth in grain boundaries at the surface of smooth specimen of 316 stainless steel in high temperature creep (in Japanese), *Journal of the Society of Materials Science Japan*, 31, pp. 505–509. https://www.jstage.jst.go.jp/article/jsms1963/31/344/31_344_505/_pdf/-char/ja

Ohtani, R., S. Nakayama, and T. Taira, 1984, Applicability of creep J-integral to microcrack propagation of creep in 304 stainless steel (in Japanese), *Journal of the Society of Materials Science Japan*, 33, pp. 590–595. https://www.jstage.jst.go.jp/article/jsms1963/33/368/33_368_590/_pdf/-char/ja

Ohtani R., T. Kinami, and H. Sakamoto, 1986, Small crack propagation in high temperature creep-fatigue of 304 stainless steel, *Transaction of*

JSME, Ser. A, 52, pp. 1824–1830. https://www.jstage.jst.go.jp/article/kikaia1979/52/480/52_480_1824/_pdf/-char/ja

Ohtani R., and T. Kitamura, 1987, Numerical simulation of microstructurally short-crack propagation in creep, *JSME International Journal, Ser. A*, 30, pp. 1741–1749.

Ohtani R., T. Kitamura, H. Murayama, and N. Tada, 1988, Simulation of initiation and early propagation of creep-fatigue small cracks based on the model of random fracture resistance of grain boundaries (in Japanese), *Transaction of JSME, Ser. A*, 54, pp. 1312–1316. https://www.jstage.jst.go.jp/article/kikaia1979/54/503/54_503_1312/_pdf/-char/ja

Ohtani R., T. Kitamura, Y. Kuriyama, M. Abe, and H. Miki, 1990, Surface small crack initiation and growth of heat resisting alloys in creep-fatigue (in Japanese), *Journal of the Society of Materials Science Japan*, 39, pp. 529–535. https://www.jstage.jst.go.jp/article/jsms1963/39/440/39_440_529/_pdf/-char/ja

Ohtani R., T. Kitamura, and H. Miki, 1991, Crack propagation of oxide dispersion strengthened superalloy Inconel MA754 in creep-fatigue (in Japanese), *Journal of the Society of Materials Science Japan*, 40, pp. 1297–1302. https://www.jstage.jst.go.jp/article/jsms1963/40/457/40_457_1297/_pdf/-char/ja

Okada M., M. Tsutsumi, T. Kitamura, and R. Ohtani, 1998, Initiation and growth of small cracks in directionally solidified Mar-M 247 under creep-fatigue, Part I: Effect of microstructure, *Journal of Fatigue and Fracture of Engineering Materials and Structures*, 21, pp. 741–750.

Okazaki M., T. Endoh, T. Yoda, and T. Koizumi, 1987, Surface small crack growth behavior of SUS 304 stainless steel in low cycle fatigue under creep-fatigue condition at elevated temperature (in Japanese), *Journal of the Society of Materials Science Japan*, 36, pp. 1232–1238. https://www.jstage.jst.go.jp/article/jsms1963/36/410/36_410_1232/_pdf/-char/ja

Okazaki M., T. Yoda, and T. Endo, 1989, Surface small crack growth behavior in low-cycle fatigue at elevated temperature and application limit of macroscopic crack growth law, *Nuclear Engineering Design*, 111, pp. 123–134.

Sakurai S., S. Usami, S. Umezawa and H.Miyata, 1986, A remaining life prediction method on the basis of micro-crack initiation and growth behavior under creep-fatigue in plain specimen of 316 stainless steel (in Japanese), *Journal of the Society of Materials Science Japan*, 35, pp. 170–175. https://www.jstage.jst.go.jp/article/jsms1963/35/389/35_389_170/_pdf/-char/ja

Tada N., T. Kitamura, and R. Ohtani, 1990, Monte Carlo simulation of creep-fatigue small cracks based on a three-dimensional model of random fracture resistance of grain boundaries (in Japanese), *Transaction of JSME, Ser. A*, 56, pp. 708–714. https://www.jstage.jst.go.jp/article/kikaia1979/56/524/56_524_708/_pdf/-char/ja

Tada N., S. Iio, T. Kitamura, and R. Ohtani, 1992, Transition behavior of creep-fatigue crack propagation from microstructurally small cracks to mechanically large crack (in Japanese), *Journal of the Society of Materials Science Japan*, 41, pp. 1165–1171. https://www.jstage.jst.go.jp/article/jsms1963/41/466/41_466_1165/_pdf/-char/ja

Tada N., W. Zhou, T. Kitamura, and R. Ohtani, 1995, Creep-fatigue intergranular fracture of inner cracking type in Type 304 stainless steel (in Japanese), *Journal of the Society of Materials Science Japan*, 44, pp. 84–89. https://www.jstage.jst.go.jp/article/jsms1963/44/496/44_496_84/_pdf/-char/ja

Zhou W., R. Ohtani, T. Kitamura, N. Tada, and A. Kosaka, 1995, Creep-fatigue intergranular fracture of inner cracking type in Type 304 stainless steel (in Japanese), *Journal of the Society of Materials Science Japan*, 44, pp. 78–83. https://www.jstage.jst.go.jp/article/jsms1963/44/496/44_496_78/_pdf/-char/ja

Index

For Product Safety Concerns and Information please contact our EU representative GPSR@taylorandfrancis.com Taylor & Francis Verlag GmbH, Kaufingerstraße 24, 80331 München, Germany